普.通.高.等.学.校
计算机教育"十二五"规划教材

计算机科学导论

(英中双语版)

AN INTRODUCTION TO COMPUTER SCIENCE
(English-Chinese Bilingual)

[英]邓辉舫(Deng Huifang)
邹才凤(Zou Caifeng)◆编著
[巴基斯坦]娜西娅·安瓦(Nazia Anwar)

人民邮电出版社
北京

图书在版编目（CIP）数据

计算机科学导论 = An Introduction to Computer Science：汉、英 /（英）邓辉舫，邹才凤，（巴基）娜西娅•安瓦（Nazia Anwar）编著. -- 北京：人民邮电出版社，2016.9（2023.9重印）
普通高等学校计算机教育"十二五"规划教材
ISBN 978-7-115-43550-7

Ⅰ. ①计… Ⅱ. ①邓… ②邹… ③娜… Ⅲ. ①计算机科学－高等学校－教材－汉、英 Ⅳ. ①TP3

中国版本图书馆CIP数据核字(2016)第215204号

内 容 提 要

本书用英汉双语写成，系统而又概要地介绍了计算机学科的定义、范畴、特点以及发展与变化的规律等。本课程的目的不是让学生学习大量具体的专业知识，而是针对所有对计算机感兴趣的学生进行入门性导引。全书由 16 章组成，分为 6 大部分：绪论，数据表示与操作，计算机硬件，计算机软件，数据组织与抽象以及高级与前沿专题（例如，物联网、云计算、大数据等），涵盖了计算机科学各个领域（包括最新领域）的最新进展、所有经典与丰富的前沿主题，以及跨学科、跨行业的应用。可以拓展读者的知识视野，能够帮助他们较全面地了解计算机科学的学科特点与发展现状，为今后深入学习计算机科学的各门专业课程打下基础。本书每一章末尾都包含本章小结，附有参考读物及一定量的练习，以帮助学生巩固已学的知识和课后进行深入的探讨。

本书既适合国内高等院校各专业用作计算机基础课教材，也可以供希望了解计算机相关领域知识的非专业读者作为入门参考。

◆ 编　著　[英] 邓辉舫（DENG Huifang）
　　　　　　　　邹才凤（ZOU Caifeng）
　　　　　[巴基] 娜西娅•安瓦（Nazia Anwar）
　　责任编辑　张　斌
　　责任印制　彭志环

◆ 人民邮电出版社出版发行　北京市丰台区成寿寺路11号
　邮编 100164　电子邮件 315@ptpress.com.cn
　网址　https://www.ptpress.com.cn
　涿州市京南印刷厂印刷

◆ 开本：787×1092　1/16
　印张：17.5　　　　　　　　2016年9月第1版
　字数：455千字　　　　　　2023年9月河北第11次印刷

定价：49.80 元

读者服务热线：(010)81055256　印装质量热线：(010)81055316
反盗版热线：(010)81055315

ABOUT THE AUTHOR

Prof. and Dr. Deng Huifang (Chinese-British) is a full Professor at South China University of Technology (SCUT, "211" and "985") in Guangzhou, China. He received his BSc and MSc degrees in China and his PhD degree from University College London (UCL). From the year of 1989 to 2004, he was studying, working and living in the UK (for more than 15 years). From the year of 2001 to 2004, he served the Phillips (UK R&D Center) and Sunrise Systems Limited as the Chief Technical Officer and Chief Scientist at Cambridge in the UK. In May 2004, SCUT publicly called for worldwide applications for the position of the Dean of Software School. He then won the "battle" and took the position. His research interests include RFID technology and applications, Internet of things, cloud computing, data science, information diffusion theory, social computing, computer simulation and modeling, service and advanced computing, massive parallel high-performance supercomputing, large-scale scientific supercomputing etc. So far, he has got over 100 papers published in English, co-chaired more than 10 International Conferences outside China, and hosted China "863" and provincial-level key research projects. He has delivered the lecture on "Introduction to Computer Science" for 10 years in English to Chinese and foreign students who are majored in computer science, finance, civil engineering, environmental engineering and so on.

作者简介

邓辉舫（Deng Huifang）是华南理工大学（"211""985"高校）计算机科学与工程学院英籍教授，博士生导师。他在中国获得理学学士和理学硕士学位，在英国伦敦大学（UCL）获得博士学位。他留英16年，曾经在英国曼彻斯特（Manchester）大学、布里斯托尔（Bristol）大学、利物浦（Liverpool）大学、伦敦大学玛丽皇后（Queen Mary）学院、伦敦大学学院（University College London）学习和工作，曾在飞利浦英国研发中心任职，并在剑桥旭日系统软件公司任技术总监和首席科学家。2004年华南理工大学全球公开招聘软件学院院长，他应聘回国任职。他的研究方向包括RFID技术与应用、物联网、云计算、大数据、信息扩散理论、社会计算、计算机模拟仿真、服务与先进计算、大规模高性能超级并行计算（量子计算、流体动力学计算）、大规模超级科学计算（分子动力学计算）等。到目前为止，他发表了100多篇英文论文，主持了国家"863"项目及多个省部级重点项目，主持过十余次国际会议。他给各专业的中国学生及外国留学生全英文讲授《计算机科学导论》近10年，积累了丰富的教学经验与心得。

Preface

Computer is constantly innovating all aspects of our lives. Computer science is a rapidly developing and expanding discipline. Computer networks link the world as a whole, and change the world we are living into a "global village". The Internet of Things enables us to find anything at any time and anywhere. The cloud computing makes us to use the computing or storing resources as conveniently as we use water and electricity every day. The data science equips us with powerful intelligence to probe the unknown world. Breakthroughs in artificial intelligence exhibited by AlphaGo has surprised us. Novel material and drug syntheses, sophisticated defense system implementation, and the life's code cracking heavily rely on supercomputers. Internet and e-commerce has changed our way of life. Even the evolution of the universe can be simulated by supercomputers. The human beings have gained huge profits by artificial intelligence, virtual reality, computer-aided design, medical diagnostics, computer animation, image & video processing, scientific computing, wearable computing, ubiquitous computing, multimedia, robotics, telecommunication and networking, internet of things, cloud computing, data science, and so on. Undoubtedly, computer has integrated into every aspect of our work and life. It has become the vital supporting tool for scientific research, business, finance, innovation of industry, national and social security, health care, public services, and so on. This book provides a concise introduction to computer science. It covers a wide range of subjects and topics, theoretical and applied, classic and state-of-the-art. It is not only suitable for students majoring in computer science, but also for those in all other majors who are interested in or wish to seek a broad and complete introduction.

Exclusive features

This book is easy for beginners to understand and helpful to stimulate students for further exploration. Several outstanding features of this book make itself unique.

This book contains a wide range of topics, varying in level of details, sophistication and difficulty. Its scope covers not only classic but also brand new hot areas, including but not limited to the topics from the classical data representation and operation, computer hardware and software, data organization and abstraction, network, algorithms, programming languages, software engineering, computational theory, to the cutting-edge hot topics, such as the fastest supercomputer, cloud computing, big data, Internet of Things, the latest video encoding and decoding methods, the popular programming languages collections, the latest operating systems, the vision of 5G wireless communications, the quantum cryptography in the network security, the latest types of databases and management tools and the powerful computing platforms for big data, and the latest progress in the artificial intelligence etc.

This book uses many figures, tables and examples to demonstrate or interpret or visualize concepts. Throughout this book, we lay more emphasis on concepts or applications rather than theoretical models.

The style of the book is rich in visual and intuitive presentation. Many figures, tables, and examples are presented in the book to help readers grasp important and sophisticated concepts with ease.

Especially, for all key technical terms or phrases, idioms, and "hard nuts" appeared in this book, the corresponding Chinese translations are inserted just beside. This unique bilingual feature would help not only Chinese students, but also international students and technicians get a better understanding of the concepts in computer science or arouse their interest in taking this course.

Outline of the Course

After an introductory chapter, the book is divided into five parts.

Part I: Data representation and operation

This part includes Chapters 2, 3, and 4. Chapter 2 discusses number systems and their conversions; how a quantity can be represented using symbols or digital pattern and what are the mutual converting rules. Chapter 3 discusses data storage and data compression. Chapter 4 discusses some primitive operations on bits.

Part II: Computer hardware

This part consists of Chapters 5 and 6. Chapter 5 gives an overview of computer hardware, explaining different computer organizations. Chapter 6 discusses how individual computers are connected to make computer networks and internets and what kind of communication models and protocols are used. In particular, this chapter explores some areas related to the Internet and its applications.

Part III: Computer software

This part includes Chapters 7, 8, 9, and 10. Chapter 7 explores operating systems. Chapter 8 shows how to design an algorithm for problems. Chapter 9 takes a journey through the landscape of contemporary programming languages. Chapter 10 provides an overall review of software engineering.

Part IV: Data organization and abstraction

Part IV contains Chapters 11 and 12. Chapter 11 discusses data structures and abstract data types, and shows how different file structures can be used for different purposes. Chapter 12 discusses databases we come across every day.

Part V: Advanced topics

This part comprises Chapters 13, 14, 15, and 16. Chapter 13 deals with some issues related to network security. Chapter 14 discusses the theory of computation. Chapter 15 is an introduction to artificial intelligence, a topic with day-to-day challenges in computer science. Chapter 16 discusses Internet of Things, cloud computing, and data science.

Acknowledgments

Great thanks are due to those who participated in each draft version's preparation of this book:

Zhifei Lai(赖志飞), Long Chen(陈龙), Hong Yao(姚宏), Xiu Wu(吴秀), Shuaishuai Yang(杨帅帅), Panpan Fan(范盼盼), Guang Xu(徐光), Hui Zhang(张慧), Youjun Liu(刘有君), Mingfei Zhao(赵明飞), Yuansheng He(何远生), Junjun Zhou(周君君), Weimin Peng(彭伟民), Wen Lai(赖雯), Yun Luo(罗云), Yuxin Liu(刘玉新), Zhipeng Fu(扶志鹏), Hao Cheng(程浩), Siyu Tang(唐思瑜), and Zhengyong Chen(陈正镛). Without their dedicated work, it is hard to imagine this book can be completed in such a state-of-the-art presentation of the contents. We also want to thank South China University of Technology for providing institutional support.

目 录

Unit 1　Introduction ········· 1
- 1.1　What Is Computer Science? ········· 1
- 1.2　Von Neumann Model ········· 1
- 1.3　Computer Components ········· 2
- 1.4　History and Development Trends ········· 3
- 1.5　Frontiers of Computer Technology ········· 5
- 1.6　Major Fields of Computer Science ········· 12
- 1.7　References and Recommended Readings ········· 13
- 1.8　Summary ········· 14
- 1.9　Practice Set ········· 14

Unit 2　Number Systems and Conversions ········· 16
- 2.1　Introduction ········· 16
- 2.2　Positional Number Systems ········· 17
- 2.3　Non-Positional Number Systems ········· 26
- 2.4　References and Recommended Readings ········· 27
- 2.5　Summary ········· 28
- 2.6　Practice Set ········· 28

Unit 3　Data Storage and Compression ········· 30
- 3.1　Bit Pattern ········· 30
- 3.2　Integer in Computer ········· 31
- 3.3　Floating-Point in Computer ········· 34
- 3.4　Data Storage ········· 36
- 3.5　Cloud Storage and Big Data Storage ········· 44
- 3.6　References and Recommended Readings ········· 48
- 3.7　Summary ········· 49
- 3.8　Practice Set ········· 51

Unit 4　Data Operations ········· 54
- 4.1　Logical Operations ········· 54
- 4.2　Arithmetic Operations ········· 59
- 4.3　Shift Operations ········· 64
- 4.4　References and Recommended Readings ········· 66

- 4.5　Summary ········· 66
- 4.6　Practice Set ········· 67

Unit 5　Computer Components ········· 70
- 5.1　Three Main Components ········· 70
- 5.2　Components Interconnection ········· 80
- 5.3　Machine Cycle ········· 82
- 5.4　Computer Architectures ········· 83
- 5.5　References and Recommended Readings ········· 84
- 5.6　Summary ········· 85
- 5.7　Practice Set ········· 86

Unit 6　Computer Networks ········· 88
- 6.1　Types of Networks ········· 88
- 6.2　TCP/IP Model ········· 91
- 6.3　Devices in Networks ········· 95
- 6.4　New Development in Networks ········· 96
- 6.5　References and Recommended Readings ········· 104
- 6.6　Summary ········· 104
- 6.7　Practice Set ········· 106

Unit 7　Operating Systems ········· 108
- 7.1　Categories ········· 109
- 7.2　Components ········· 111
- 7.3　Popular Operating Systems ········· 118
- 7.4　References and Recommended Readings ········· 121
- 7.5　Summary ········· 121
- 7.6　Practice Set ········· 122

Unit 8　Algorithm ········· 125
- 8.1　What Is Algorithm? ········· 125
- 8.2　Three Constructs ········· 126
- 8.3　How to Evaluate Algorithms ········· 127
- 8.4　Algorithm Representation ········· 128
- 8.5　Basic Algorithms ········· 129
- 8.6　Classification of Algorithm ········· 136

8.7	References and Recommended Readings ········· 140	
8.8	Summary ········· 140	
8.9	Practice Set ········· 141	

Unit 9 Programming Languages ········· 143

9.1	Development ········· 143	
9.2	Program Translation ········· 145	
9.3	Languages' Latest Ranking, Categories and Features ········· 146	
9.4	Common Concepts of Programming Languages ········· 151	
9.5	Severe Software Errors ········· 153	
9.6	References and Recommended Readings ········· 156	
9.7	Summary ········· 156	
9.8	Practice Set ········· 157	

Unit 10 Software Engineering ········· 159

10.1	Introduction ········· 159
10.2	Software Development Life Cycle ······ 160
10.3	Software Development Models ········· 165
10.4	CMMI and Software Process Management ········· 167
10.5	Importance of Documentation ········· 169
10.6	Service-Oriented Architecture ········· 170
10.7	References and Recommended Readings ········· 170
10.8	Summary ········· 171
10.9	Practice Set ········· 172

Unit 11 Data and File Structures ········· 173

11.1	Abstract Data Types ········· 173
11.2	List ········· 174
11.3	Stack ········· 177
11.4	Queue ········· 177
11.5	Tree and Graph ········· 178
11.6	File Structure ········· 180
11.7	References and Recommended Readings ········· 184
11.8	Summary ········· 184
11.9	Practice Set ········· 185

Unit 12 Databases ········· 187

12.1	Introduction ········· 187
12.2	Database Management Systems ········· 188
12.3	Database Architecture ········· 189
12.4	The History of Database Systems ········· 189
12.5	Database Model ········· 191
12.6	Relational Operations ········· 195
12.7	Databases for Big Data ········· 196
12.8	References and Recommended Readings ········· 197
12.9	Summary ········· 198
12.10	Practice Set ········· 198

Unit 13 Security ········· 200

13.1	Security Goals ········· 200
13.2	Security Threats ········· 201
13.3	Security Services ········· 206
13.4	Cryptography ········· 210
13.5	References and Recommended Readings ········· 215
13.6	Summary ········· 215
13.7	Practice Set ········· 216

Unit 14 Theory of Computation ········· 218

14.1	The Turing Machine ········· 218
14.2	Halting Problem ········· 222
14.3	Solvable Problems ········· 224
14.4	References and Recommended Readings ········· 225
14.5	Summary ········· 226
14.6	Practice Set ········· 226

Unit 15 Artificial Intelligence ········· 228

15.1	Introduction ········· 228
15.2	Knowledge Representation and Expert Systems ········· 229
15.3	Perception ········· 231
15.4	Reasoning ········· 238

15.5	Nature Inspired Computation 242	16.3	Challenges and Research Directions ... 257
15.6	References and Recommended Readings 247	16.4	IoT Applications 260
		16.5	Cloud Application Service Models 261
15.7	Summary 248	16.6	Cloud Application Deployment Models 262
15.8	Practice Set 248		
		16.7	Big Data Tools and Techniques 263

Unit 16 Internet of Things, Cloud Computing and Data Science 250

- 16.1 Introduction 250
- 16.2 Opportunities in IoT, Cloud, and Data Science 254
- 16.8 Integration of IoT, Cloud Computing, and Big Data 264
- 16.9 References and Recommended Readings 265
- 16.10 Summary 266
- 16.11 Practice Set 267

Unit 1 Introduction

This book provides a tour of computer science【计算机科学】. It presents the latest research about what computers do and how they do it, from top to bottom and outside to inside. A computer system is an accumulation of many different elements, combined to generate a whole that is far more than the sum of its parts. This book explores a wide range of topics. It aims to provide the foundational understanding of computer science, and establishes a comprehensive overview of this realm.

1.1 What Is Computer Science?

Computer science is a discipline that involves the theoretical foundations【理论基础】 of computation and their applications in computer systems. It is difficult to list a complete range of computer research areas due to their rapid pace of innovation. This discipline contains many sub-fields, such as programming methodology and languages【程序设计方法学和语言】, algorithm and data structure【算法和数据结构】, computational complexity theory【计算复杂性理论】, computer graphics【计算机图形学】, operating systems【操作系统】, software engineering【软件工程】, database systems【数据库系统】, artificial intelligence【人工智能】, computer architecture【计算机系统结构】, computer networking and communication【计算机网络与通信】, parallel computation【并行计算】, and so on.

Computer science is a relatively young discipline that begins in the 1940s. It spans theory and practice. It can be seen as a science of problem solving. It has strong associations with other disciplines. Many problems in engineering, science, medicine, agriculture, business, and other fields can be solved effectively with computers. The cross-disciplinary knowledge【交叉学科知识】 is necessary in order to find a solution.

1.2 Von Neumann Model

The von Neumann model 【冯·诺依曼模型】 is a design architecture for an electronic digital computer【电子数字计算机】. It consists of an arithmetic logic unit【算术逻辑单元】, a control unit【控制器】, a memory【内存】, an external storage【外部存储器】, and input【输入】 and output【输出】 mechanisms (Figure 1.1).

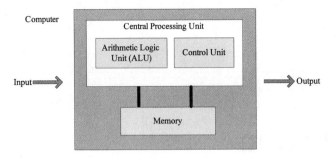

Figure 1.1　The von Neumann model

Computers based on the Turing machine【图灵机】 store data【数据】 in their memory. Around 1944-1945, John von Neumann proposed that programs【程序】 should also be stored in the memory. Both data and programs in the von Neumann model have the same logical format. They are stored as binary【二进制】 patterns in memory—a sequence of 0s and 1s. A program in the von Neumann model contains a finite number of instructions【指令】 that are executed sequentially【按顺序执行】.

The bus【总线】 is shared【共享】 between the program memory and data memory, leading to the von Neumann bottleneck【瓶颈】. It has often been considered an issue due to the limited throughput【吞吐量】 between the central processing unit (CPU)【中央处理单元（器）】 and memory, because of the single bus that can only access one of the two classes of memory at a time. When the CPU is required to execute minimal processing on large amounts of data, it is forced to wait for needed data to be transferred to or from memory, which seriously limits the processing speed.

1.3　Computer Components

A computing system consists of hardware【硬件】, software【软件】, and data【数据】. Computer hardware is the collection of physical parts of a computer, such as a monitor【显示器】, a graphic card【显卡】, a hard disk drive (HDD)【硬盘驱动器】, a sound card【声卡】, memory, a motherboard【主板】, a mouse【鼠标】, a keyboard【键盘】, and so on. Computer software is a collection of computer programs, procedures, and documentation. Data is the distinct information with which a computer system deals.

1.3.1　Computer hardware

Computer hardware refers to the components that you can physically touch. There are many parts of computer hardware that can be installed inside, or connected to the outside, of a computer. The external hardware includes a flat-panel【平板显示器】, a printer【打印机】, a projector【投影仪】, a scanner【扫描仪】, a speaker【扬声器】, a keyboard, a mouse, a flash memory【闪存】, and so on. The internal hardware includes CPU, a multi-core processor【多核处理器】, a motherboard, a network card【网卡】, a modem【调制解调器】, a sound card, a video card【显卡】, a drive (e.g. Blue-Ray【蓝光】, CD-ROM【只读光盘驱动器】, DVD【数字通用光盘】, floppy drive【软盘驱动器】, hard disk drive (HDD)【硬盘驱动器】), a solid-state disk (SSD)【固态硬盘】, and so on.

1.3.2 Computer software

Computer software is any set of programs that a computer carries out【执行】. The program consists of a sequence of instructions. Each instruction operates on one or more data items. These instructions might be internal system commands【系统命令】, or responses to external input. Software informs the various hardware components what to do and how to interact with each other.

There are two main types of software: system software【系统软件】 and application software【应用软件】. System software is used to run the hardware, while application software is used to perform other tasks. The main system software includes operating systems【操作系统】and drivers【驱动程序】. The main application software includes games, media players, word processors, anti-virus programs【杀毒软件】, and so on. Software is usually written in high-level programming languages【高级编程语言】, which are translated into the machine language【机器语言】 by a compiler【编译器】 or an interpreter【解释器】. Software may also be written in a low-level assembly language【汇编语言】.

1.3.3 Data

Data is any sequence of symbols given meaning by specific actions of interpretation【解释】. In daily life, we usually use digits that can take one of ten states (0 to 9). However, data stored in a computer generally use only two states (0 and 1). Some forms of data (text, image, audio, and video) cannot be stored in a computer directly, and needs to be converted into the binary【二进制】 form (0s and 1s).

Data can be organized in many types of data structures【数据结构】, including arrays【数组】, lists【列表】, graphs【图】, and objects【对象】. Modern high performance data persistence【数据持久性】 technologies rely on massively parallel distributed data processing【并行分布式数据处理】, such as Apache Hadoop. In such systems, the data is distributed across multiple computers and can be processed on different computers at the same time.

1.4 History and Development Trends

Computer science has undergone a rapid development since its birth. As the field of computer science has emerged, new directions of research and applications have been created and combined with classical discoveries in a continuous cycle of growth and revitalization.

1.4.1 Computer history

Life without a computer is unimaginable. The history of computer is an attractive story. Computers were not always the brilliant fast machines empowering us to obtain lots of knowledge. Actually, the first computer was very different from the recent computers. Generally speaking, history of computers can be divided into three periods.

1. Mechanical machines

In the early 17th century, a French mathematician called Blaise Pascal invented Pascaline, which is a mechanical calculator【机械计算器】 performing addition【加法】 and subtraction【减法】. In the late 17th century, Gottfried Leibniz, a German mathematician, invented Leibniz' Wheel. In the early 19th century, Joseph-Marie Jacquard invented Jacquard loom, which adopted the idea of storage and programming at the first time. In the 1820s, Charles Babbage invented Analytical Engine. In 1890, Herman Hollerith invented a programmable machine that could read and sort data on punched cards automatically.

2. Electronic computers

Between 1930 and 1950, several scientists contributed to the evolution of computer technology who could be considered the true early pioneers of computer science. John Vincent Atanasoff and his assistant Clifford Berry invented the ABC (Atanasoff Berry Computer) to solve a system of linear equations. It encoded information electrically. In the 1930s, a huge computer called Mark I, was built under the direction of Howard Aiken at Harvard University. In England, Alan Turing invented a computer called Colossus to break the German Enigma code. In 1946, John Mauchly and J. Presper Eckert invented ENIAC (Electronic Numerical Integrator and Calculator), the first totally electronic computer【电子计算机】. In 1950, EDVAC, the first computer to implement the stored program concept based on von Neumann's ideas was built at the University of Pennsylvania.

3. Computers after 1950

Between 1950 and 1959, computers were bulky and utilized vacuum tubes【电子管】 as electronic switches. Between 1959 and 1965, transistors【晶体管】 replaced vacuum tubes. Then the size and the cost of the transistorized computers were dramatically reduced. From 1965 to 1975, the appearance of the integrated circuit【集成电路】 further reduced the size and cost of computers. Between 1975 and 1985, microcomputers appeared. The Altair 8800, the first desktop calculator【台式计算机】, was invented in 1975. Between 1985 and 1995, some advanced computer technology appeared, such as Clusters【集群】, Vector Processors【向量处理器】, workstations【工作站】, minicomputers【小型计算机】, laptops【笔记本电脑】 and palmtop computers【掌上电脑】, and so on. After 1995, high-performance computers (HPC)【高性能计算机】 obtained a great advancement, such as supercomputers【超级计算机】, many-cores【多核】 personal computers【个人电脑】, graphics processing unit (GPU)【图形处理单元（器）】, general-purpose graphics processing unit (GPGPU)【通用计算图形处理单元（器）】, and so on.

1.4.2 Development trends

In the future, computer technology will be characterized by high performance, miniaturization, network, popularization, intelligence, and humanization.

1. Lightweight microcomputer

The lightweight microcomputer【微型计算机】 with small size, low price, powerful function, and high reliability will be popular.

- A Pocket PC【掌上电脑】 is a handheld device【手持设备】 that can be used to process e-mail, play games, exchange messages, browse the Web, and so on.
- A laptop computer【笔记本电脑】 has most of the same components into a single unit as a

desktop computer, such as a keyboard, a pointing device, a display, and speakers. Laptop computers have become increasingly popular because they are becoming smaller, lighter, cheaper, and more powerful, and their screens are becoming smaller and of better quality.

- A smart phone【智能手机】 is based on a mobile operating system. The early smart phones added the functions of a personal digital assistant (PDA)【个人数字助理】, a portable media player, a low-end compact digital camera, and a pocket video camera. Many modern smart phones also contain high-resolution touch screens【触摸屏】, web browsers, global positioning system (GPS)【全球定位系统】 navigation【导航】 units, and so on.
- An iPad is a kind of tablet computers【平板电脑】 built on Apple's iOS operating system. An iPad includes the functions of video players, cameras, music players, web browsers, emails, games, GPS navigations, social services, etc.

2. High performance computer

Development of powerful super computers with high speed, high performance, and capability of processing large and complex problems is also the definite trend.

- A parallel computer【并行计算机】 has a set of processors that work simultaneously【同时运行】. Parallel computers use multiple computational resources to solve large problems that can often be divided into smaller ones. From smart phones, to large supercomputers and web sites, to multi-core CPUs and GPUs, parallel processing is ubiquitous in modern computing.
- A graphics processing unit (GPU)【图形处理单元（器）】 is a specialized electronic circuit used to rapidly handle memory to accelerate the processing of images. GPU computing utilizes a GPU together with a CPU to accelerate business, scientific and engineering applications. GPU + CPU is a powerful union because GPUs consist of thousands of small, efficient cores designed for parallel computing, and CPUs consist of a few cores optimized for serial computing【串行计算】. The compute unified device architecture (CUDA)【统一计算设备架构】 is a parallel computing platform and an application programming interface (API)【应用程序编程接口】 model invented by NVIDIA. The CUDA-enabled GPU can be used for general purpose computing. The CUDA parallel computing platform provides a few simple C and C++ extensions to express data and task parallelism.
- A supercomputer【超级计算机】 has high-level processing capacity that makes it possible to calculate problems at ultra-high speed. Supercomputers can be used for highly calculation-intensive tasks such as quantum physics【量子物理】, weather forecasting, climate research, oil and gas exploration, etc.

1.5 Frontiers of Computer Technology

The frontiers of computer technology include supercomputers, cloud computing【云计算】, big data【大数据】, Internet of Things【物联网】, mobile computing【移动计算】, quantum computers【量子计算机】, biocomputers【生物计算机】, virtualization【虚拟化】, service-oriented architecture (SOA)【面向服务的体系结构】, etc.

1.5.1 Supercomputer

A supercomputer is a computer that performs calculation at the currently highest operational rate. It is typically used for scientific and engineering applications that deal with very large databases or do large amounts of computation (or both). Figure 1.2 shows that supercomputing has increasingly become the cornerstone of modern society.

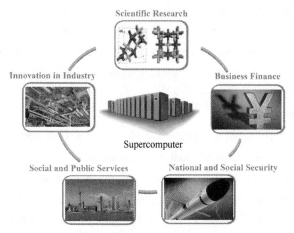

Figure 1.2　Supercomputing has increasingly become the cornerstone of modern society

Supercomputers play an important role in the field of computational science. The stages of supercomputer application include simulation【仿真】 and modeling【建模】, weather forecasting, meteorology【气象学】, geological exploration【地质勘探】, cryptography【密码学】 (encryption【加密】and decryption【解密】), virtual reality【虚拟现实】, artificial intelligence【人工智能】, nuclear weapons【核武器】 development, precise guide missile【引导导弹】, long-range attack【远程攻击】, quantum simulation【量子模拟】, novel drug synthesis and discovery【新型药物的合成和发现】, novel material synthesis【新材料的合成】, space explorations【太空探索】, data mining【数据挖掘】, business intelligence, etc. Supercomputers can be applied to calculate the structures and properties of chemical compounds【化合物】, biological macromolecules【生物大分子】, polymers【聚合物】, and crystals【晶体】. High performance supercomputers are widely employed in physical simulations, such as simulation of airplanes in wind tunnels【风洞】, simulation of the detonation【爆炸】 of nuclear weapons, and research into nuclear fusion【核聚变】.

According to the 47th edition of the TOP 500 list of the world's most powerful supercomputers announced in June 2016, China maintained its No. 1 ranking, but with a new system built entirely using processors designed and made in China. Sunway TaihuLight 【神威·太湖之光】 is the new No. 1 system with 93 petaflop/s (quadrillions of calculations per second【每秒1000万亿次运算】) on the LINPACK benchmark.

Developed by the National Research Center of Parallel Computer Engineering & Technology (NRCPCET) and installed at the National Supercomputing Center in Wuxi, Sunway TaihuLight displaces Tianhe-2 (Milkyway-2)【天河二号】, an Intel-based Chinese supercomputer that has retained the No. 1 position on the past six TOP 500 lists. Table 1.1 shows the top 10 sites for June 2016.

Table 1.1 TOP 10 Sites for June 2016

(Courtesy of http://www.top500.org/lists/2016/06/)

Rank	Supercomputer	Site	Manufacturer	Country
1	Sunway TaihuLight - Sunway MP P, Sunway SW26010 260C 1.45GHz, Sunway	National Supercomputing Ce nter in Wuxi	NRCPC	China
2	Tianhe-2 (MilkyWay-2) - TH- IVB-FEP Cluster, Intel Xeon E5-2692 12C 2.200GHz, TH Express-2, Intel Xeon Phi 31S1P	National Super Computer Center in Guangzhou	NUDT	China
3	Titan - Cray XK7, Opteron 6274 16C 2.200GHz, Cray Gemini interconnect, NVIDIA K20x	DOE/SC/Oak RidgeNational Laboratory	Cray Inc.	United States
4	Sequoia - BlueGene/Q, Power BQC 16C 1.60 GHz, Custom	DOE/NNSA/LLNL	IBM	United States
5	K computer, SPARC64 VIIIfx 2.0GHz, Tofu interconnect	RIKEN Advanced Institute for Computational Science (AICS)	Fujitsu	Japan
6	Mira - BlueGene/Q, Power BQC 16C 1.60GHz, Custom	DOE/SC/Argonne National Laboratory	IBM	United States
7	Trinity - Cray XC40, Xeon E5-2698v3 16C 2.3GHz, Aries interconnect	DOE/NNSA/LANL/SNL	Cray Inc.	United States
8	Piz Daint - Cray XC30, Xeon E5-2670 8C 2.600GHz, Aries interconnect , NVIDIA K20x	Swiss National Supercomputing Centre (CSCS)	Cray Inc.	Switzerland
9	Hazel Hen - Cray XC40, Xeon E5-2680v3 12C 2.5GHz, Aries interconnect	HLRS - Höchstleistungs rechenzentrum Stuttgart	Cray Inc.	Germany
10	Shaheen II - Cray XC40, Xeon E5-2698v3 16C 2.3GHz, Aries interconnect	King Abdullah University of Science and Technology	Cray Inc.	Saudi Arabia

Sunway TaihuLight, with 10,649,600 computing cores comprising 40,960 nodes, is twice as fast and three times as efficient as Tianhe-2, which posted a LINPACK performance of 33.86 petaflop/s. Table 1.2 shows the technique data of Sunway TaihuLight.

Table 1.2 The technique data of Sunway TaihuLight

(Courtesy of http://www.top500.org/system/178764)

Site	National Supercomputing Center in Wuxi
CPU	Shenwei-64
Manufacturer	NRCPC
Node Processor Cores	256 CPEs (Computing Processing Elements) plus 4 MPEs (Management Processing Elements)
Node Peak Performance	3.06 TFlop/s
Clock Frequency	1.45 GHz
Nodes	40,960
Cores per node	60 cores
Total Cores	10,649,600
Linpack Performance (Rmax)	93,014.6 TFlop/s (or 93 PFlop/s) (TFlop = 10^{12})
Theoretical Peak (Rpeak)	125,436 TFlop/s (or 125.4 PFlop/s) (TFlop = 10^{12})
Nmax	12,288,000
Power	15,371.00 kW (or 15.3 MW)
Memory	1,310,720 GB (or 32 GB*40,960 nodes = 1.31 PB)
Processor	Sunway SW26010 260C 1.45GHz
Interconnect	Sunway
Operating System	Sunway Raise OS 2.0.5

Sunway TaihuLight is characterized by 4 supers, namely, super high speed, super high capacity, super large size, and super heavy energy consumption (Figure 1.3).

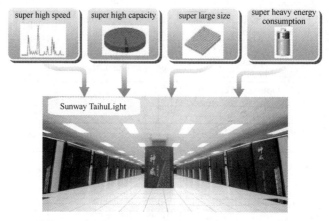

Figure 1.3　The characteristics of Sunway TaihuLight

Exponential growth of supercomputing power was recorded by the TOP 500 list. Figure 1.4 shows the achieved performance development of supercomputers. Figure 1.5 shows the projected performance development of supercomputers. $EFlop = 10^{18}$, $PFlop = 10^{15}$, $TFlop = 10^{12}$.

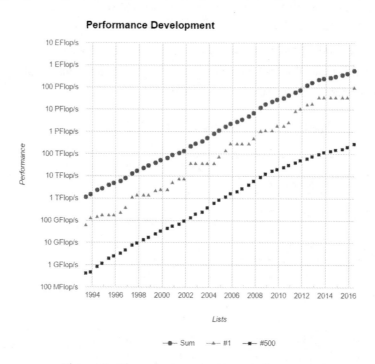

Figure 1.4　The performance development of supercomputers

(Courtesy of http://www.top500.org/statistics/perfdevel/)

A new breakthrough by Japanese and German scientists, by using the almost 83,000 processors of one of the world's most powerful supercomputers, was able to mimic just one percent of one second's worth of human brain activity—and even that took 40 minutes.

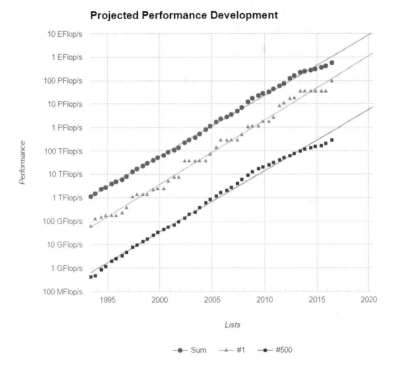

Figure 1.5　The projected performance development of supercomputers

(Courtesy of http://www.top500.org/statistics/perfdevel/)

1.5.2　Cloud computing

Cloud computing is a model for enabling ubiquitous, convenient, on-demand【按需随选】network access to a shared pool of configurable computing resources (e.g., networks, servers, storage, applications, and services), according to the NIST (National Institute of Standards and Technology) definition. Cloud computing has become one of the most powerful and economical paradigms in terms of technology, architecture, and IT services for performing complex computations in large-scale scientific and business applications. A large number of organizations have been shifting to cloud computing environment. The benefits of cloud computing include virtualized resources【虚拟化的资源】, elasticity【自由伸缩性】, scalability【可扩展性】, flexibility【灵活性】 of services, on-demand delivery of resources, efficiency, parallel processing, cost saving, and a lot more.

Cloud computing is a form of utility computing【效用计算】 where IT resources are delivered on a Pay-As-You-Use【使用才支付】 basis, just as electricity is charged based on usage. Cloud computing has transformed the IT industry into more economical and flexible model for the sharing of resources. Using cloud computing, large organizations can process their large-scale jobs more quickly, since using hundred servers per hour outweighs the benefits than using one server for hundred hours. The paradigm of cloud computing has unprecedentedly revolutionized the IT industry.

In cloud computing, users can ubiquitously run applications, and store/access their data on an off-site, location-transparent【位置透明】, centrally managed, shared platform. The resources provided by cloud are usually run on virtual machines【虚拟机】paralleled with the physical infrastructure. There are many types of public cloud computing, such as Infrastructure as a service (IaaS)【基础设施即服务】, Platform as a service (PaaS)【平台即服务】, Software as a service (SaaS)【软件即服务】, Data as a service (DaaS)【数据即服务】, and so on.

A supercomputer consists of the best processors, rapid memory, specially designed components, and elaborate cooling mechanisms, so it is not easy to scale a supercomputer. On the contrary, the distributed cloud computing is much more affordable and scalable. The processing power of cloud computing grows as additional servers (with their processors) are added to the network. On the other hand, supercomputers have the advantage of sending data through fast connections, while cloud computing architecture sends data through slower networks.

1.5.3 Big Data

Data, structured or unstructured, produced by organizations, Internet of Things (IoT)【物联网】and social media is growing tremendously. This fast and sudden growth of big data requires fast development of big data technologies through virtualization platforms. Big data refers to the large increase in the volumes of data that needed advanced technologies and techniques to capture, store, process, distribute, manage and analyze the information. The terms volume, variety, and velocity were initially presented by Gartner to describe challenges of big data. Gantz and Reinsel proposed that the elements of big data are characterized by 4Vs, namely, volume【容量大】, variety【种类多】, velocity【流量大，处理速度快】, and value【价值高】.

Volume refers to the continuously expanding magnitude of all types of data produced from various sources, such as IoT, social media, multimedia, and a lot more. Variety refers to the various structures of data gathered through IoT, social networks, smart devices, or sensors. Velocity refers to the rate of data generation and speed of data transfer and analysis. Value is very crucial characteristic of big data. It is the process of rapidly determining hidden patterns or information from various types of big data.

IT organizations want to take advantage of cloud computing architecture to support big data projects. Cloud computing provides enterprises cost-effective, flexible access to big data's huge volumes of information. Meanwhile big data on the cloud generates plentiful on-demand computing resources. Both cloud computing and big data technologies will continue to evolve and congregate in the future.

1.5.4 Internet of Things

The Internet of Things, or IoT, is a network of items, each embedded with sensors【传感器】, which are connected to the Internet. IoT is defined as the worldwide dynamic global autonomous network of interconnected physical or virtual things, devices or objects based on standard network protocols【网络协议】. IoT refers to an emerging paradigm consisting of a continuum of uniquely addressable things communicating with one another to form a worldwide dynamic network.

Nowadays, many smart devices【智能设备】are connected to the Internet; smart buildings have multiple sensors to protect them from disasters or accidents and to save energy; and smart devices have supported various fields including healthcare services, business, education, and industry. Huge volumes of data and information are generated by these devices, which are connected directly or indirectly via

Internet.

Cloud based on platforms and applications helps to connect objects in IoT at any place and at any time. Therefore, the front end to access IoT is the cloud. Cloud provides the needed infrastructure【基础设施】, computation powers, storage and applications to interact with smart devices in real-time.

The huge amounts of data produced from the connections of devices in IoT can only be captured, processed, stored and transformed into valuable information through clouds. Clouds provide physical and virtual systems, applications and tools to efficiently and intelligently process and analyze big data and IoT.

1.5.5　Mobile computing

Mobile computing is a technology allows transmission of data, via a computer or any other wireless device without having to be connected to a fixed physical link. Mobile computing involves mobile communication【移动通信】, mobile hardware, and mobile software. Mobile communication issues include ad-hoc【点对点】 and infrastructure networks【基础设施网络】, communication properties, protocols【协议】, data formats and concrete technologies, etc. Hardware includes mobile devices or components. Mobile software handles the characteristics and requirements of mobile applications. One of the most important characteristics of mobile computing is portability. Mobile computing can use a computing device even when being mobile or changing location.

1.5.6　Quantum computer

A quantum computer【量子计算机】 makes direct use of quantum mechanical phenomena, such as superposition【叠加】 and entanglement【纠缠】, to perform calculations. Quantum computers are different from digital computers based on transistors. Quantum computation makes use of quantum properties to represent data and perform operations on data, while digital computers require data to be encoded into binary digits. Quantum computation uses quantum bits【量子比特（位）】, which can be in superpositions of states. Quantum computers have the ability to be in multiple states simultaneously. Large-scale quantum computers may theoretically be able to solve certain problems asymptotically faster than any conventional computer by using the best algorithms, such as integer factorization【整数分解】 using Shor's algorithm.

1.5.7　Biocomputer

Biocomputers【生物计算机】 use biologically derived materials, such as DNA and proteins【蛋白质】, to perform computational functions. Currently, biocomputers have various functional capabilities, such as operations of binary logic and mathematical calculations. Three typical types of biocomputers include biochemical computers, biomechanical computers, and bioelectronic computers. In the future, biocomputers can obtain a long-term development by the expanding new science of nanobiotechnology【纳米生物技术】.

1.5.8　Virtualization

Virtualization【虚拟化】 is the simulation of the software or hardware upon which other software runs. It creates virtualized components including hardware platforms, operating systems (OS), storage devices, and network devices, etc. Based on virtualization technology, the IT environment will be able to manage itself based on perceived activity, and utility computing. Virtualization is aimed to centralize administrative tasks, improve scalability and overall hardware-resource utilization. Virtualization reduces overhead costs due to parallelism of operating systems on a single central processing unit. Using virtualization, an enterprise can better control updates of the operating system and applications without disrupting users.

1.5.9　Service-oriented architecture (SOA)

A service-oriented architecture (SOA)【面向服务的体系结构】 is a set of methodologies for designing software in the form of interoperable services【可互操作的服务】. These services can provide the functionalities of large software applications that can be reused whenever needed. Service-orientation needs loosely coupled【松耦合】 services between operating systems and other technologies that underlie applications. In SOA architecture, services are independent of any other vendor, product, service, or technology. The architectural pattern of SOA consists of application components that provide services to other components via a communication protocol, mostly over a network.

1.6　Major Fields of Computer Science

Computer science is the field to study the theoretical foundations of computation, together with their implementation and applications in computer systems. The major fields of computer science are given as follows:

（1）Mathematical foundations. The research contents include mathematical logic【数理逻辑】, set theory【集合论】, number theory【数论】, graph theory【图论】, category theory【范畴论】, numerical analysis【数值分析】, information theory【信息论】, and so on.

（2）Theory of computation【计算理论】. The research contents include automata theory【自动机理论】, computability theory【可计算性理论】, computational complexity theory【计算复杂性理论】, quantum computing theory, and so on.

（3）Algorithms【算法】 and data structures【数据结构】. The research contents include analysis of algorithms, algorithm design, computational geometry【计算几何】, and so on.

（4）System architecture. The research contents include digital logic【数字逻辑】, memory systems, computer architecture, computer organization, operating systems, and so on.

（5）Concurrent【并发】, parallel【并行】, and distributed systems【分布式系统】. The research contents include multiprocessing【多重处理】, grid computing【网格计算】, concurrency control, and so on.

（6）Programming languages and compilers【编译器】. The research contents include parsers【解析器】, interpreters, procedural programming, object-oriented programming, functional programming, logic programming, and so on.

（7）Software engineering【软件工程】. The research contents include requirements analysis, software design, computer programming, formal methods, software testing, software development, and so on.

（8）Databases【数据库】. The research contents include data mining, relational databases【关系型数据库】, online analytical processing (OLAP)【联机分析处理】, distributed and parallel database systems, database security【安全】and privacy【机密性】, and so on.

（9）Artificial intelligence【人工智能】. The research contents include automated reasoning【自动推理】, computational linguistics【计算语言学】, computer vision【计算机视觉】, evolutionary computation【进化计算】, machine learning【机器学习】, neural networks【神经网络】, natural language processing【自然语言处理】, robotics【机器人学】, and so on.

（10）Computer graphics【计算机图形学】 and visual computing. The research contents include visualization【可视化】, image processing【图像处理】, advanced geometric modeling【几何建模】, mainstream rendering【主流渲染】 techniques, computer animation【计算机动画】, and so on.

（11）Human computer interaction【人机交互】. The research contents include computer accessibility【可访问性】, user interfaces【用户接口】, wearable computing【可穿戴计算】, ubiquitous computing【普适计算】, virtual reality, multimedia【多媒体】, and so on.

（12）Telecommunication【通信】 and networking. The research contents include routing【路由】, network topology【网络拓扑结构】, cryptography【密码学】, security and privacy, and so on.

（13）Scientific computing. The research contents include artificial life【人工生命】, bioinformatics【生物信息学】, cognitive science【认知科学】, computational chemistry【计算化学】, computational neuroscience【计算神经科学】, computational physics【计算物理学】, numerical algorithms【数值算法】, symbolic mathematics【符号数学】, and so on.

1.7 References and Recommended Readings

For more details about the subjects discussed in this chapter, the following books are recommended:

- Behrouz A Forouzan, Firouz Mosharraf. Foundations of Computer Science, second edition, London: Thomson Learning, 2008.
- Schneider G M, Gersting J L. Invitation to Computer Science, Boston, MA: Course Technology, 2004.
- Dale N, Lewis J. Computer Science Illuminated, Sudbury, MA: Jones and Bartlett, 2004.
- Patt Y, Patel S. Introduction to Computing Systems, New York: McGraw-Hill, 2004.
- Allen B Tucker, Peter Wegner. Computer Science Handbook, second edition, Chapman and Hall/CRC, 2004.
- ACM/IEEE-CS Joint Task Force. Computing Curricula 2001, Computer Science Volume. ACM and IEEE Computer Society, December 2001. (http://www.acm.org/sigcse/cc2001).
- http://en.wikipedia.org/wiki/Main_Page.
- http://www.nvidia.co.uk/object/tesla-high-performance-computing-uk.html.

- http://www.top500.org/.
- http://www.nvidia.com/object/cuda_home_new.html.
- Armbrust M, Fox A, Griffith R, et al, A view of cloud computing. Communications of the ACM, 2010, 53(4), 50-58.
- Gantz J, Reinsel D. Extracting value from chaos, IDC iView, 2011,1-12.
- Gandomi A Haider M. Beyond the hype: Big data concepts, methods, and analytics. International Journal of Information Management, 2015,35(2), 137-144.
- Gubbi J, Buyya R, Marusic S, et al. Internet of Things (IoT): A vision, architectural elements, and future directions. Future Generation Computer Systems, 2013,29(7), 1645-1660.

1.8 Summary

- Computer science deals with the study and the science of the theoretical foundations of information and computation and their implementation and application in a wide variety of computer systems.
- The von Neumann architecture has four subsystems: memory, arithmetic logic unit, control unit, and input/output. The von Neumann model states that the program instructions and data must be stored in main memory.
- We can conceptualize a computer as being made up of three important components: computer hardware, computer software, and data.
- In general, the history of computing and computers can be classified into three periods: the period of mechanical machines (before 1930), the period of electronic computers (1930-1950), and the period after generations (1950-present).
- In the future the development of computer technology will represent miniaturization, high performance, network, popularization, intelligence and humanization.
- The frontiers of computer technology include supercomputer, cloud computing, big data, Internet of Things, mobile computing, quantum computer, biocomputer, virtualization, service-oriented architecture (SOA), and so on.
- The major fields of computer science include mathematical foundations, theory of computation, algorithms and data structures, system architecture, parallel and distributed systems, programming languages and compilers, software engineering, databases, artificial intelligence, computer graphics, human computer interaction, telecommunication and networking, scientific computing, and so on.

1.9 Practice Set

1. Define a computer model based on the von Neumann architecture.
2. What is the role of a program in a computer based on the von Neumann model?

3. The _____ model is the basis for today's computers.
   a. Leibnitz   b. von Neumann   c. Pascal   d. Charles Babbage
4. In a computer, the _____ subsystem stores data and programs.
   a. ALU   b. input/output   c. memory   d. control unit
5. In a computer, the _____ subsystem performs calculations and logical operations.
   a. ALU   b. input/output   c. memory   d. control unit
6. In a computer, the _____ subsystem serves as a manager of the other subsystems.
   a. ALU   b. input/output   c. memory   d. control unit
7. According to the von Neumann model, _____ stored in memory.
   a. only data is   b. only programs are
   c. data and programs are   d. none of the above
8. According to the von Neumann model, can the hard disk of today be used as input or output? Explain.
9. Analyze and summarize the development tendency of computer technology?
10. What is supercomputer? Which supercomputer is the new No. 1 system with 93 petaflop/s (quadrillions of calculations per second) on the LINPACK benchmark in June 2016?
11. What is cloud computing? What are the characteristics of cloud computing?
12. What are the characteristics of big data?
13. What is the definition of IoT?
14. What are the major research fields of computer science?

Unit 2
Number Systems and Conversions

This chapter is a prelude to Chapters 3 and 4. In Chapter 3 we show how data is stored inside the computer. In Chapter 4 we show how logic and arithmetic operations【逻辑和算术运算】are performed on data. This chapter is a preparation for understanding the contents of Chapters 3 and 4. You are probably already familiar with many of the concepts about binary numbers described in this chapter, but you might not realize that you know them! All number systems follow the same rules; it's just a matter of grasping the underlying concepts involved and applying them in a new base.

2.1 Introduction

Number is the basic concept in mathematics. There are various kinds of numbers, such as natural numbers【自然数】, negative numbers【负数】, rational numbers【有理数】, irrational numbers【无理数】, etc. Now, let's give a brief introduction to those numbers.

First, what is a number? A number is defined as a mathematical object used to count, measure, and label. Then, how to separate these numbers? A natural number is the number 0 or any number produced by repeatedly adding 1 to this number【自然数是数字 0 或所有在这个数字基础上连续加 1 得到的数字】. Natural numbers are used for counting. A negative number is less than 0, and its sign【符号】is opposite to a positive number. For example, -5 is a number that is five less than 0. A rational number is the number that can be represented as one integer divided by another, but an irrational number cannot be represented as a ratio of integers, i.e. a fraction.

Actually, a number system【计数系统；数制】is a series of rules about how a number of specific value can be represented using various symbols【计数系统是一系列描述如何将代表特定数值的数字用不同符号表示的规则】. We can represent a number in different systems. For example, the two numbers $(2A)_{16}$ and $(52)_8$ have the same value: $(42)_{10}$, and we know that their representations are different.

In the past, we had utilized various different number systems; all of them can be approximately categorized into two groups: positional number systems【按位计数系统；位置化的计数系统】and non-positional number systems【非按位计数系统；非位置化的计数系统】. We mainly discuss these two number systems, but our main goal is to discuss the positional number systems, and give examples of non-positional systems in order to develop a deep understanding of number systems.

2.2 Positional Number Systems

Actually, the positional number system often appears in our daily life. In a positional number system, the value a number represents is determined by the position a symbol occupies in this number. In this system, we can represent a number as follows:

$$\pm(a_{k-1}\cdots a_2 a_1 a_0 . a_{-1} a_{-2} \cdots a_{-l})_r$$

It has the value of:

$$t = \pm(a_{k-1} \times r^{k-1} + \cdots + a_1 \times r^1 + a_0 \times r^0 + a_{-1} \times r^{-1} + a_{-2} \times r^{-2} + \cdots + a_{-l} \times r^{-l})$$

where a is the set of symbols, r is the base (or radix)【基数】.

We use the base of a number system to identify the number of digits used in the number system, and the digits always begin with 0 and continue through one less than the base. Usually, there are 2 digits (0 and 1) in binary number【二进制数】system, 8 digits (0 through 7) in octal number【八进制数】system, and 10 digits (0 through 9) in decimal number【十进制数】system. The base also uses the position of digits to indicate meaning. When we add 1 to the last digit in the number system, we need to have a carry to the digit position to the left.

Maximum value

Sometimes we need to know the maximum value of a decimal integer that can be represented by k digits for base r. The answer is $r^k - 1$. For example, if $k = 5$, $r = 10$, then the maximum value is 99999.

2.2.1 The decimal system (base 10)

In our daily life, we always come into contact with some decimal numbers. In this section, let's get to know more about the decimal system【十进制系统】. The word decimal comes from the Latin root decem (ten). In this system the base $r = 10$ and we use ten symbols:

A = {0, 1, 2, 3, 4, 5, 6, 7, 8, 9}

The symbols in this system are often called decimal digits or just digits. In the decimal system, we can represent a number as follows:

$$\pm(a_{k-1}\cdots a_2 a_1 a_0 . a_{-1} a_{-2} \cdots a_{-l})_{10}$$

If not leading to some ambiguity, we often omit the parentheses, the base, and the plus sign for simplicity. For example, we write the number as 163.06.

Integers

All of us are very familiar with integers【整数】in the decimal system because we often use them in our daily life. We represent an integer as N, and the value is calculated as:

$$N = \pm(a_{k-1} \times 10^{k-1} + a_{k-2} \times 10^{k-2} + \cdots + a_2 \times 10^2 + a_1 \times 10^1 + a_0 \times 10^0)$$

where a_i is a digit, $r = 10$ is the base, and k is the number of digits.

Example 2.1

Now, let's pay attention to this example, in which we can see the place values【位值】for the integer +375 in the decimal system.

```
Place values        10²          10¹          10⁰
Number               3            7            5
Value          +  3×10²   +   7×10¹   +   5×10⁰   =  N
```

Reals

A real【实数】 is a number with a fractional part【小数部分】 which is also familiar to us. For example, we use this system to show dollars and cents (92.35). We can represent a real as $\pm(a_{k-1}\cdots a_2 a_1 a_0 \cdot a_{-1} a_{-2} \cdots a_{-l})$. We calculate its value as follows:

Integral part .   Fractional part
$$R = \pm((a_{k-1} \times 10^{k-1} + \cdots + a_1 \times 10^1 + a_0 \times 10^0) + (a_{-1} \times 10^{-1} + \cdots + a_{-l} \times 10^{-l}))$$

where a_i is a digit, $r = 10$ is the base, k is the number of digits in the integral part 【整数部分】, and l is the number of digits in the fractional part. The decimal point in the representation is used to separate the integral part from the fractional part.

Example 2.2

The following shows the place values for the real number in decimal system +91.52.

```
Place values        10¹          10⁰          10⁻¹         10⁻²
Number               9            1            5            2
Value          +  9×10   +   1×1   +   5×0.1  +  2×0.01    = R
```

2.2.2 The binary system (base 2)

The binary system 【二进制系统】 is fundamental to computer science because of the physical features in computers. We must get to know the base-2 (binary) number system if we want to find out how the program runs in a computer. Then it will be easier for us to understand other number systems that are powers of 2, such as base 8 (octal), and base 16 (hexadecimal). In the binary system the base r = 2 and we use only two symbols, A = {0, 1}. We often regard these symbols in this number system as binary digits 【二进制数字】 or bits 【比特, 位】.

Bit is the smallest unit of the binary system. However, in the real applications, one prefers to use byte (B), which equals 8 bits, to measure the capacity of a storage device, or a memory chip or a binary number stored in the computer. If the capacity or the number is very large, one has to prefix a letter to the bye (B) to save space. There are many prefix letters, such as KB (KiloByte), MB (MegaByte), GB (GigaByte), TB (TeraByte), PB (PetaByte), EB (ExaByte), ZB (ZettaByte), YB (YottaByte), BB (BrontoByte), NB (NonaByte), DB (DoggaByte), and CB(CorydonByte) etc. The definition of each prefix is listed in the table 2.1.

Table 2.1 The definition of each prefix letter

Unit (Prefix)		Bytes (B) in binary		Roughly (≈) Bytes (B) in decimal	Exact bits (Notes)
B	Byte【字节】	2^0	1	10^0	8
KB	Kilo【千】	2^{10}	1,024	10^3	8,192
MB	Mega【兆】	2^{20}	1,048,576	10^6	8,388,608
GB	Giga【吉】	2^{30}	1,073,741,824	10^9	8,589,934,592
TB	Tera【太】	2^{40}	1,099,511,627,776	10^{12}	8,796,093,022,208 (Mass data【海量数据】)
PB	Peta【拍】	2^{50}	1,125,899,906,842,624	10^{15}	9,007,199,254,740,992
EB	Exa【艾】	2^{60}	1,152,921,504,606,846,976	10^{18}	9,223,372,036,854,775,808
ZB	Zetta【泽】	2^{70}	1,180,591,620,717,411,303,424	10^{21}	9,444,732,965,739,290,427,392 (Big data【大数据】)
YB	Yotta【尧】	2^{80}	1024ZB	10^{24}	8192Z
BB	Bronto【博】	2^{90}	1024YB	10^{27}	8192Y
NB	Nona【诺】	2^{100}	1024BB	10^{30}	8192B
DB	Dogga【多】	2^{110}	1024NB	10^{33}	8192N
CB	Corydon【秾】	2^{120}	1024DB	10^{36}	8192D

Same as the decimal system, we can represent an integer as $\pm(a_{k-1}\cdots a_1 a_0)_2$, and we calculate its value as follows:

$$N = \pm(a_{k-1}\times 2^{k-1} + a_{k-2}\times 2^{k-2} + \cdots + a_2\times 2^2 + a_1\times 2^1 + a_0\times 2^0)$$

where a_i is a digit, $r = 2$ is the base, and k is the number of bits.

Similar to the real in decimal system, a real in binary system is also a number with an optional fractional part, we can represent a real as $\pm(a_{k-1}\cdots a_2 a_1 a_0 . a_{-1} a_{-2} \cdots a_{-l})$, and we calculate its value as follows:

Integral part . Fractional part

$$R = \pm((a_{k-1}\times 2^{k-1} + \cdots + a_1\times 2^1 + a_0\times 2^0) + (a_{-1}\times 2^{-1} + \cdots + a_{-l}\times 2^{-l}))$$

Where a_i is a digit, $r = 2$ is the base, k is the number of bits in the integral part, and l is the number of digits in the fractional part. The decimal point we use in our representation separates the fractional part from the integral part.

Example 2.3

The following shows that the number $(110.01)_2$ in binary is equivalent to the number 6.25 in decimal.

Place values	2^2	2^1	2^0	2^{-1}	2^{-2}	
Number	1	1	0	0	1	
Value	1×2^2 +	1×2^1 +	0×2^0 +	0×2^{-1} +	1×2^{-2}	= **R**

Therefore, the equivalent decimal number is $R = 4 + 2 + 0 + 0 + 0.25 = 6.25$.

2.2.3 The hexadecimal system (base 16)

Because of the physical features of computers, the numbers are stored in binary form, but it is not convenient for us to write their representations【因为计算机的一些物理特性，数字都是以二进制方式存储的，但二进制表达式不方便我们书写】. So, can we try to use decimal system? Unfortunately, it is not convenient to store data in computers because there is no obvious relationship between the number of bits in binary and the number of decimal digits. In order to solve this problem, we utilize other number systems: hexadecimal【十六进制】 and octal, which makes better interaction between people and computers.

Now, let's first discuss the hexadecimal system. The word hexadecimal is derived from the Greek root hex (six) and the Latin root decem (ten). In the hexadecimal system the base r is equal to 16 and we need sixteen symbols to represent a hexadecimal number. The set of symbols is {0, 1, 2, 3, 4, 5, 6, 7, 8, 9, A, B, C, D, E, F} and the symbols A, B, C, D, E, F represent the values of 10, 11, 12, 13, 14, 15 respectively. We regard these symbols in hexadecimal system as hexadecimal digits【十六进制数字】.

We can represent an integer as $\pm(a_{k-1}\cdots a_1 a_0)_{16}$, and we calculate its value as follow:

$$N = \pm(a_{k-1} \times 16^{k-1} + a_{k-2} \times 16^{k-2} + \cdots + a_2 \times 16^2 + a_1 \times 16^1 + a_0 \times 16^0)$$

where a_i is a digit, $r = 16$ is the base, and k is the number of digits.

Example 2.4

The following example shows how to utilize the place value to calculate the value of number $(3BD)_{16}$.

Place values	16^2	16^1	16^0	
Number	3	B	D	
Value	3×16^2 +	11×16^1 +	13×16^0	$= N$

Therefore, the equivalent decimal number is $N = 768+176+13 = 957$.

2.2.4 The octal system (base 8)

In this section, we discuss the octal system【八进制系统】. The word octal comes from the Latin root octo (eight). In this system the base r is equal to 8 and we need eight symbols to represent an octal number. The set of symbols in this system is {0, 1, 2, 3, 4, 5, 6, 7}, and we regard the symbols in this system as octal digits.

We can represent an integer as $\pm(a_{k-1}\cdots a_1 a_0)_8$. The value is calculated as:

$$N = \pm(a_{k-1} \times 8^{k-1} + a_{k-2} \times 8^{k-2} + \cdots + a_2 \times 8^2 + a_1 \times 8^1 + a_0 \times 8^0)$$

Where a_i is a digit, $r = 8$ is the base, and k is the number of digits.

Example 2.5

The following example shows that the number $(2147)_8$ in octal is equivalent to $(1127)_{10}$ in decimal.

Place values	8^3	8^2	8^1	8^0	
Number	2	1	4	7	
Value	2×8^3 +	1×8^2 +	4×8^1 +	7×8^0	$= N$

2.2.5 Summary of the four positional systems

Table 2.2 represents a summary of the four positional number systems discussed in this chapter.

Table 2.2 Summary of the four positional number systems

Number system (base)	Symbols	Examples
Decimal (10)	0,1,2,3,4,5,6,7,8,9	2223.89
Binary (2)	0,1	$(1110.01)_2$
Octal (8)	0,1,2,3,4,5,6,7	$(7116.02)_8$
Hexadecimal (16)	0,1,2,3,4,5,6,7,8,9,A,B,C,D,E,F	$(FC2.2E)_{16}$

Table 2.3 shows the representation of the numbers 0 to 15 in different number systems.

Table 2.3 Comparison of numerals in the four systems

Decimal	Hexadecimal	Octal	Binary
0	0	0	0000
1	1	1	0001
2	2	2	0010
3	3	3	0011
4	4	4	0100
5	5	5	0101
6	6	6	0110
7	7	7	0111
8	8	10	1000
9	9	11	1001
10	A	12	1010
11	B	13	1011
12	C	14	1100
13	D	15	1101
14	E	16	1110
15	F	17	1111

2.2.6 Conversion

Now, we have talked about how to represent a number in the specified number system. In this section, our aim is to get an understanding of conversion between different number systems. Since people are more familiar with the decimal system than the other systems. Therefore, we first show how to covert from any base to decimal, and then show how to convert from decimal to any base. At last, we show how to convert from binary to hexadecimal or octal, and vice versa.

1. Conversion from other bases to decimal

The task of converting to decimal number from another base is comparatively easier. We first determine the place value of each digit, then multiply each digit with its corresponding place value. Finally, add the results of the products to get the equivalent number in decimal【我们先确定每个数字的位值，然后将每个数字和相应的位值相乘，最后将相乘的结果相加得到对应的十进制数】. Figure 2.1 shows the idea of converting from any base into decimal.

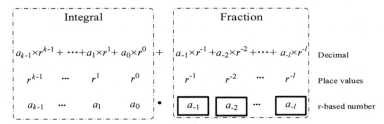

Figure 2.1　Converting from other bases into decimal

Now, let's take a look at the following examples.

Example 2.6

We can get more details from the following example of converting the binary number $(101.01)_2$ to decimal: $(101.01)_2 = 5.25$.

Binary	1	0	1	0	1	
Place values	2^2	2^1	2^0	2^{-1}	2^{-2}	
Decimal	4	+ 0	+ 1	+ 0	+ 0.25	= **5.25**

Example 2.7

This is another example of converting the hexadecimal number $(2B.48)_{16}$ to decimal.

Hexadecimal	2	B	4	8	
Place values	16^1	16^0	16^{-1}	16^{-2}	
Decimal	32	+ 11	+ 0.25	+ 0.03125	= **43.28125**

2. Conversion from decimal to other bases

In the process of converting decimal number to other bases, we divide the specified decimal number by the base of target number system, then, we can get a quotient【商】 and a remainder【余数】. The reminder becomes the next digit (from right to left) of new representation in the target number system, and the quotient becomes the new number to be divided until the quotient equals zero【余数成为目标计数系统中新的表达式的下一个数字（从右至左），商继续被除直到等于零】. The source decimal numbers may have both integral part and fractional part. Therefore, converting a decimal number to other bases is actually two-step process.

Converting the integral part

Converting the integral part of the specified number is a simple process; we use the skill of repetitive division【累积相除】 to convert the integral part. Figure 2.2 shows the framework of the conversion process.

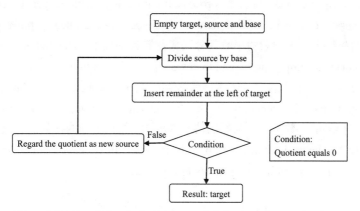

Figure 2.2　Converting the integral part of decimal number to other bases

Now, let's discuss some details about this type of conversion. For simplicity, we regard the integral part of the specified decimal number as the source, and the converted representation as target. The initial work is to make an empty target, and then repeatedly dividing the source in order to get corresponding quotient and remainder【最初先建立空的目标表达式，然后将待转换的数重复地除以基数获得相应的商和余数】. The quotient becomes new source participating in the repetitive division process until it equals zero, and the remainder is inserted to the left of the target【商成为新的待转换的数，并参与下一次累积相除过程直到商等于零，并且余数插入目标表达式的左边】. Figure 2.3 shows the framework of the conversion process.

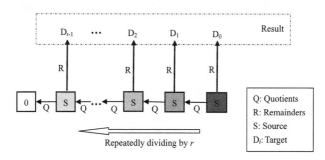

Figure 2.3　Converting the integral part of a decimal number to other bases

The following example shows the details of the conversion process.

Example 2.8

The following shows how to convert 47 in decimal to binary. We start with the number in decimal, we move to the left while dividing repeatedly by 2, and continuously finding the quotients and the remainders as we go. The result is $47 = (101111)_2$.

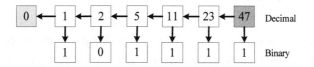

Converting the fractional part

Similar to the process of converting integral part, we use the skill of repetitive multiplication 【累积相乘】 to convert the fractional part. The process of converting fractional part is also very simple. In the conversion process, we regard the fractional part of the specified decimal number as the source【在转换过程中，我们将特定十进制数的小数部分当成待转换的数】. The initial work of converting fractional part is to make an empty target, and then we repeatedly multiply the source by the base【转换小数部分时先建立空的目标表达式，然后将待转换的数重复地乘以基】. The fractional part of multiplied source becomes new source, while the integral part of multiplied source becomes the right digit of target【相乘得到的小数部分成为新的待转换的数，相乘得到的整数部分成为目标表达式右边的数字】. Figure 2.4 shows the framework of converting the fractional part in decimal to other bases. We can get the details of each multiplying process in Figure 2.5.

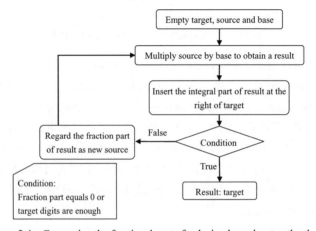

Figure 2.4 Converting the fractional part of a decimal number to other bases

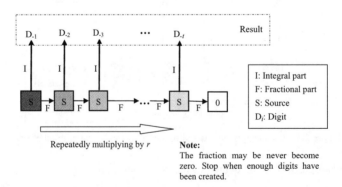

Figure 2.5 Converting the fractional part of a decimal number to other bases

Example 2.9

We have obtained the main idea of converting decimal number to other bases, now let's discuss the question how to convert the decimal number 0.375 to binary.

Solution

We discover that the number $0.375 = (0.011)_2$ has no integral part, so we omit the process of converting the integral part. The following shows details of the conversion process.

Example 2.10

This is another example how to convert 0.374 to octal, and we take the maximum of four digits. So the result is 0.374 = $(0.2773)_8$. In the process of converting, the base equals 8. The following shows the process of converting a decimal fraction to binary.

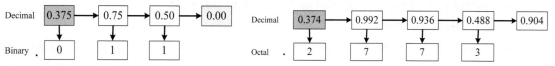

Example 2.11

This example shows how to convert 195.4 in decimal to hexadecimal having only one digit to the right of the decimal point. We get the result: $(C3.6)_{16}$. The following shows the details of the conversion process.

Solution

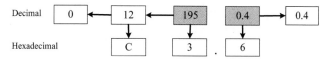

3. Binary-hexadecimal conversion

Now, how can we solve the problem of binary-hexadecimal conversion? Actually, this problem is easier than before. Since there is an interesting relationship between the two bases: four bits is one hexadecimal digit【4 比特等于 1 位 16 进制数字】. Figure 2.6 vividly shows this relationship.

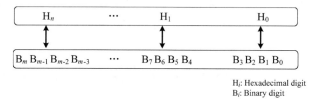

Figure 2.6　Converting hexadecimal to binary and binary to hexadecimal

Example 2.12

Let's discuss the conversion process between the binary number $(10110101001)_2$ and its corresponding hexadecimal equivalent. We first arrange the binary number in 4-bit patterns 101 1010 1001. We know $(1001)_2$ equals $(9)_{16}$, $(1010)_2$ equals $(A)_{16}$, and $(101)_2$ equals $(5)_{16}$. So we can obtain the corresponding hexadecimal equivalent of $(10110101001)_2$: $(5A9)_{16}$.

4. Binary-octal conversion

Based on the idea of binary-hexadecimal conversion, we can easily solve the problem of binary-octal conversion. The relationship between the two bases is: three bits in binary is one digit in octal【三比特等于一位八进制数字】. Figure 2.7 vividly shows this relationship.

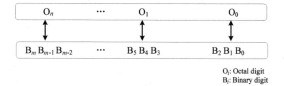

Figure 2.7　Converting octal to binary and binary to octal

Example 2.13

What is the octal equivalent of the binary number $(111101011)_2$?

Solution

The bits of binary number $(111101011)_2$ are separated into three groups, and the equivalent octal digit of each group of three bits is obtained. Finally, we get the result: $(753)_8$.

5. Octal-hexadecimal conversion

Because of the binary-hexadecimal relationship and binary-octal relationship, the task of directly converting a number from octal to hexadecimal and vice versa is easier than converting through binary system. The specified number is first translated into equivalent binary number, then the binary number is converted to the representation in the target number system. Figure 2.8 shows an example.

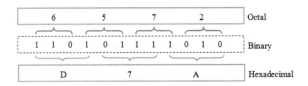

Figure 2.8 Converting octal to hexadecimal and hexadecimal to octal

The process of converting an octal number to corresponding hexadecimal one can be regarded as a combination of octal-binary conversion and binary-hexadecimal conversion. We first find the binary representation of the specified octal number, and then translate the binary representation into the hexadecimal representation. The process of converting a hexadecimal number to octal is the inverse.

2.3 Non-Positional Number Systems

Actually, modern computers do not use non-positional number systems【非按位计数系统；位置化的计数系统】, and the computer data is stored using binary. However, we need to know a little about non-positional system. In this section, we will give a brief discussion about non-positional systems. Non-positional number systems still have some limited roles in our life. Similar to positional number system, there are fixed number of symbols in non-positional number system and every symbol has a fixed value. However, the way of calculating the value of a number in non-positional number system is very special. Unlike the numbers in positional number system, the position of symbols in a number normally has no relation to its value. In this system, a number is represented as:

$$a_{k-1}\cdots a_2 a_1 a_0 . a_{-1} a_{-2} \cdots a_{-l}$$

It has the value of:

$$n = \pm((\underbrace{a_{k-1}+\cdots+a_1+a_0}_{\text{Integral part}})+(\underbrace{a_{-1}+a_{-2}+\cdots+a_{-l}}_{\text{Fractional part}}))$$

Example 2.14

Among the non-positional number systems, roman numerals are most familiar to us. This number system is based on a set of seven different symbols S = {I, V, X, L, C, D, M}. The symbols of Roman numerals and their values are shown in Table 2.4.

Table 2.4   Values of symbols in the Roman numerals

Symbol	I	V	X	L	C	D	M
Value	1	5	10	50	100	500	1000

In order to calculate the value of a roman numeral, we must follow its specific rules:

(1) When the same symbols are written together, the values are added【当相同的符号写在一起时，将它们的数值相加】, such as III = 1 + 1 + 1 = 3.

(2) When a symbol with a larger value is placed on the left of a symbol with smaller value, the values are added【当一个数值较大的符号放在一个数值较小的符号的左边，将它们的数值相加】, such as VI = 5 + 1 = 6.

(3) When a symbol with a larger value is placed on the right of a symbol with a smaller value, the smaller value is subtracted from the larger one【当一个数值较大的符号放在一个数值较小的符号的右边，将较大的数值减去较小的数值】, such as IX = 10 − 1.

(4) When a bar is placed over a symbol, it means multiplying the value with 1000【当一个符号上面有一个横杠时，表示将这个符号代表的数值乘以 1000】, such as $\bar{I}$ = 1000.

The following shows some Roman numerals and their values.

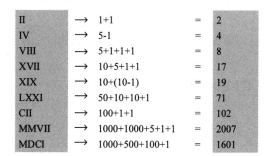

2.4   References and Recommended Readings

For more details about the subject discussed in this chapter, the following books are recommended.

- Dale N, Lewis J. Computer Science Illuminated, Sudbury, MA: Jones and Bartlett, 2004.
- Forouzan B, Mosharraf F. Foundations of Computer Science, London: Cengage Learning EMEA, 2008.
- Forouzan B A, Fegan S C. Foundations of Computer Science: From Data Manipulation to Theory of Computation, Beijing: Thomson Learning, 2003.
- Hall R W. A Course in Multicultural Mathematics. PRIMUS, 17(3), 209-227, 2007.
- Mano M M. Computer System Architecture, Upper Saddle River, NJ: Prentice Hall, 1993.
- Null L, Lobur J. Computer Organization and Architecture, Sudbury, MA: Jones and Bartlett, 2003.
- Stallings W. Computer Organization and Architecture: Designing for Performance, Pearson Education, 2015.
- Stein W. Elementary Number Theory, 2005.

2.5　Summary

- Numbers consist of several symbols in the number system.
- Number systems can be approximately classified into two systems: positional number system and non-positional number system.
- In a positional number system, the position of a symbol in the number presentation has its own specific value, which is called a place value.
- In a non-positional number system, every symbol have specific values, but the position of a symbol in the number presentation has no relationship with its value.
- In a decimal system, the base equals 10 and there are 10 symbols. The symbols are regarded as decimal digits.
- In a binary system, the base equals 2 and there are just 2 symbols. The symbols are regarded as binary digits.
- In a hexadecimal system, the base equals 16 and there are 16 symbols. The symbols are regarded as hexadecimal digits.
- In an octal system, the base equals 8 and there are 8 symbols. The symbols are regarded as octal digits.
- In the process of converting a number in any number system to a decimal number, we multiply each digit with its corresponding place value, and add them together to obtain the result.
- When we convert a real, we need two steps: one is converting the integral part of the real; another is converting the fraction part of the real. The former uses repetitive division, the latter uses repetitive multiplication.
- Because four digits in a binary system can be represented by one digit in a hexadecimal system, we can use this relationship to solve the problems of conversion between a binary system and a hexadecimal system.
- Similar to the conversion between a binary system and a hexadecimal system, as three digits in a binary system can be represented by one digit in an octal system, we can make full use of this relationship to solve the problems of conversion between a binary system and an octal system.
- In the process of conversion between an octal system and a hexadecimal system, we can make full use of the binary-octal relationship and binary-hexadecimal relationship to realize our goal.

2.6　Practice Set

1. What is number system?
2. Explain some basic concepts about a number system.
3. What is the difference between positional and non-positional number system?
4. How many bits in a binary system are represented by one digit in an octal system?

5. Explain the main idea of conversion between an octal system and a hexadecimal system.
6. The base of the decimal number system is _____ .
   a. 8   b. 2   c. 16   d. 10
7. Which of the following representations is incorrect? _____ .
   a. $(11010010)_2$ b. $(944)_8$ c. $(4FB)_{16}$ d. 3552
8. Which of the following is equivalent to 13 in decimal? _____ .
   a. $(1101)_2$ b. $(17)_8$ c. $(F)_{16}$ d. none of the above
9. Convert the following binary numbers to decimal: _____ .
   a. $(11001110)_2$ b. $(101101)_2$ c. $(10110.101)_2$ d. $(100111.111)_2$
10. Convert the following hexadecimal numbers to decimal: _____ .
    a. $(AC2)_{16}$ b. $(123)_{16}$ c. $(ABB)_{16}$ d. $(35E.E1)_{16}$
11. Convert the following octal numbers to decimal: _____ .
    a. $(437)_8$ b. $(2271)_8$ c. $(223.7)_8$ d. $(471.21)_8$
12. Convert the following decimal numbers to binary: _____ .
    a. 1314   b. 21   c. 314.02   d. 14.56
13. Convert the following decimal numbers to octal: _____ .
    a. 1756   b. 346   c. 121.4   d. 48.8
14. Convert the following decimal numbers to hexadecimal: _____ .
    a. 717   b. 121   c. 212.13   d. 316.5
15. Convert the following octal numbers to hexadecimal: _____ .
    a. $(514)_8$ b. $(411)_8$ c. $(13.7)_8$ d. $(1256)_8$
16. Convert the following decimal numbers into binary: _____ .
    a. $4\,^5/_8$ b. $2\,^3/_{32}$ c. $14\,^{13}/_{64}$ d. $22\,^5/_{128}$
17. Find the maximum value of an integer in each of the following cases: _____ .
    a. b=10, k=10   b. b=2, k=12   c. b=8, k=8   d. b=16, k=7
18. Write the decimal equivalents of the following Roman numerals: _____ .
    a. XXVII   b. XV   c. MCLVII   d. VLIII

Unit 3
Data Storage and Compression

To learn how a computer processes data, it is necessary to understand the nature of data. This chapter introduces various types of data, and focuses on how to store and compress data in a computer. Besides, cloud storage and massive data storage are introduced briefly.

3.1 Bit Pattern

Not so long ago, computers were used to do scientific calculations and display human-readable text. But now they are actually multimedia devices, dealing with many different types of data, including numbers, text, audio, images, and video.

For efficiency, all aforementioned【之前提到的】 data types have a uniform representation inside the computer, which is based on a binary system【二进制系统】, called a bit pattern【位模式/位组】 or binary representation (Figure 3.1).

```
10111111
```
Figure 3.1 A bit pattern

In fact, there have been computers based on decimal number system【十进制系统】 or other number systems. But these computers are too complicated, expensive, and less stable for every day use. Intuitively, if a single logic unit represents less possible values, the computer will be more stable. For an electronic signal, the binary system provides only two possible states: low or high, corresponding to 0 or 1 respectively. An electronic switch can be easily represented by two stable states of electronic signal. These sates can be defined as on and off represented by 1 and 0 respectively【一个电子开关，可以很容易地用电子信号的两种稳定状态来表示。即分别由 1 和 0 代表的开和关】.

To represent more than two things (on and off), one bit is not enough, so we need a sequence of bits. In general, m bits are capable of representing 2^m things because m bits can make 2^m combinations of 0 and 1.

Computer memory 【内存】 has no idea what type of data a stored bit pattern represents. The memory can store a piece of data with the same pattern ignoring its data type【内存可以使用相同的模式，存储一份数据，而忽略该数据的数据类型】, as shown in Table 3.1.

Table 3.1 Storage of different data types

Data types	Program	Memory
A number '38'	Math routine	00100110
A character '&'	Text editor	00100110
Part of an image 头	Image recorder	00100110
Part of a song 〰	Music recorder	00100110
Part of a film ◉	Video recorder	00100110

3.2 Integer in Computer

Integers represent numbers which have no fractional part. As discussed in Chapter 2, each bit pattern can be treated as a binary number, and each binary number can be converted to the corresponding decimal integer. It's natural to use a bit pattern to represent integers. As Figure 3.2 shows, a fixed-point representation 【定点表示法】 is the method used for storing integers in binary format. The decimal point is assumed at the right of the least significant (rightmost) bit without any gap.

Figure 3.2 Fixed point representation of integers

To make more efficient use of computer memory, unsigned integer【无符号整数】 and signed integer【有符号整数】 have been developed as two categories of integer representation. Three distinct ways have been developed to represent signed integers (Figure 3.3).

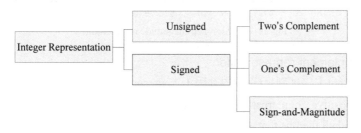

Figure 3.3 Classification of integers

3.2.1 Unsigned representation

Unsigned representation is used to represent positive integers and zero. Storing unsigned integers is a straightforward process, which is almost the same as the procedures shown in Chapter 2. After the number is changed to binary of m bits, the computer adds $m-n$ 0s to the left when n is smaller than m【在数被改变为 m 位的二进制表示后，计算机增加 $m-n$ 个 0 到它的左侧，其中 n 小于 m】. For instance, an 8-bit memory location uses unsigned representation 00001111 to store 15. Note that the left leading 0s are essential.

3.2.2 Signed representation

Unlike positive integers【正整数】, negative integers【负整数】are represented with a minus sign in front of them. To represent negative integers in binary number systems, signed representation has been developed. Inspired by the binary system, a sign can be indicated by one bit which has exactly two values. All of the above three signed representation formats use 0 for positive sign and 1 for negative sign as the leftmost bit【上述三种有符号整数的表示格式，都是用最左边的位作为符号，其中 0 表示正号，1 表示负号】. For a positive integer, its convention is the same as an unsigned representation, and the sign bit equals 0. So the range of positive integers represented by an unsigned representation is different from positive integers represented by a signed representation. The latter's range is half of the former one. The three signed representations to encode negative integers are slightly different with the signed representation to encode positive integers.

1. Sign-and-magnitude

Sign-and-magnitude【原码】 representation is simple. We can transform its magnitude part to decimal value directly. Figure 3.4 shows the range of sign-and-magnitude representation with 5-bit allocation. Since only seven bits can be used to represent the magnitude part, this range is divided into two halves: 00000 to 01111 and 10000 to 11111. The former represents positive integers, and the latter represents negative integers. You may notice there are two zeros: +0 and −0. We often ignore the negative zero entirely, but within the computer, it may cause unnecessary complexity. It's easy to conclude that an *m*-bit location can represent numbers from $-(2^{m-1}-1)$ to $+(2^{m-1}-1)$.

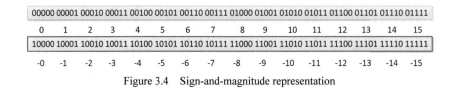

Figure 3.4 Sign-and-magnitude representation

2. One's complement representation

One's complement【二进制反码/一的补码】 representation of a negative binary integer is the complement of its positive counterpart. For a number, its complement is obtained by changing all the bits that are 0 to 1 and all the bits that are 1 to 0. The range of signed integers using one's complement in a computer is the same as the range of sign-and-magnitude integers【在计算机中，使用反码的有符号整数的范围，与原码表示的整数的范围，是相同的】(Figure 3.5). There are also two representations of 0 in this format, denoted by -0 (negative zero) and +0 (positive 0).

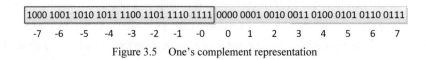

Figure 3.5 One's complement representation

For example, we can store −268 in a 16-bit memory location using one's complement representation. The number can be changed to binary 0000000100001100 with seven leading 0s added.

The sign is negative, so each bit is inverted to get the number in 1's complement. The result is 1111111011110011 (Figure 3.6).

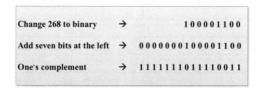

Figure 3.6　−268 in one's complement format

For one's complement representation, an 8-bit memory location cannot store a positive integer larger than 255 or a negative integer less than -255, otherwise overflow would occur. Because of negative zero and end-around borrow (which may be produced by one's complement subtraction), one's complement representation is not commonly used now.

3. Two's complement representation

In fact, two's complement 【二进制补码】 is the most common method used by computer to represent and store a signed integer with *m*-bit. An unsigned integer of (0 to 2^m-1) using two's complement representation also has two available subranges (Figure 3.7). Unlike the previous two methods for signed integers, two's complement has only one representation of zero, and 1000 is a special case whose leftmost bit 1 not only represents a negative sign, but also contributes to the decimal value. So the range of numbers that *m*-bit location can store using two's complement is -2^{m-1} to + $(2^{m-1}-1)$.

```
1000 1001 1010 1011 1100 1101 1110 1111  0000 0001 0010 0011 0100 0101 0110 0111
 -8   -7   -6   -5   -4   -3   -2   -1    0    1    2    3    4    5    6    7
```

Figure 3.7　Two's complement representation

For positive integer, the conversion from decimal to binary is enough, just like before. Note that the left padding 0s are still needed if bits are not enough. If the integer is negative, the operation includes two steps. Firstly, we invert bits from the right until the current bit is 1; then, we invert the other bits【如果这个整数是负的，则操作包括两个步骤。首先，从右边开始按位取反，直到当前位为1；然后，我们对其他位取反】. The two's complement of an integer can also be completed by adding 1 to the result after taking the one's complement.

For example, two's complement representation to store −5 in a 4-bit memory location is shown in Figure 3.8.

Figure 3.8　−5 in two's complement format

Table 3.2 shows the differences among all these integer representations. In this table, we assume that *m* is 4, so the memory location can store only 4 bits. Note that the interpretation is different for negative integer, while it is the same for positive integer.

Table 3.2 Summary of integer representations

Contents of Memory	Unsigned	Sign-and-magnitude	One's complement	Two's complement
0000	0	+0	+0	+0
0001	1	+1	+1	+1
0010	2	+2	+2	+2
0011	3	+3	+3	+3
0100	4	+4	+4	+4
0101	5	+5	+5	+5
0110	6	+6	+6	+6
0111	7	+7	+7	+7
1000	8	−0	−7	−8
1001	9	−1	−6	−7
1010	10	−2	−5	−6
1011	11	−3	−4	−5
1100	12	−4	−3	−4
1101	13	−5	−2	−3
1110	14	−6	−1	−2
1111	15	−7	−0	−1

3.3 Floating-Point in Computer

A number which consists of an integral part and a fractional part is called a real number【一个数，如果由一个整数部分和一个小数部分组成，称为实数】. Although we can use fixed-point representation to represent a real number, however, the result is hard to be accurate enough because the digits in integral part and fractional part are both fixed. The solution is to use floating-point representation 【浮点表示法】.

Floating-point representation adopts scientific notation, with only one non-zero digit to the left of the decimal point 【浮点表示法采用科学记数法，其在小数点的左侧，只有一个非零数字】. In the binary system, this digit can only be 1. As shown in Figure 3.9, the process of converting a real (101100010.01 in the following example) to scientific notation is called normalization【归一化/规范化】. After normalization, only sign, exponent and mantissa【尾数】need to be stored. Note that the exponent base 2, leftmost bit 1, and decimal point in the mantissa are implicit.

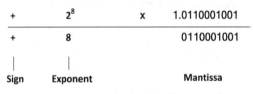

Figure 3.9 Normalization of a binary number

Converting the fraction【小数】part of a real number to binary is similar to changing a natural number from base 10 to base 2, but instead of dividing, it multiplys the fraction part by the base. The whole number part of the result is the first binary digit to the right of the point. Next, the fractional part of the previous result is multiplied by base 2. The steps are repeated until the fractional part is zero or until an infinite repeating pattern is recognized【重复这些步骤，直到小数部分是零，或直到被识别出无限重复的模式】. For example, we convert .25 into its binary representation to illustrate the method.

.25*2=0.50 ⇒ .50*2=1.00

Thus, .25 in decimal is .01 in binary. Here is another example.

.815*2=1.630 ⇒ .630*2=1.260 ⇒ .260*2=0.520 ⇒
.520*2=1.040 ⇒ .040*2=0.080 ⇒ .080*2=0.160 …

Thus, .815 is 110100… in binary.

Now let's go through the entire conversion process by converting 30.75 from decimal to binary. Firstly we convert 30.

$$2\overline{)30} \Rightarrow 2\overline{)15} \Rightarrow 2\overline{)7} \Rightarrow 2\overline{)3} \Rightarrow 2\overline{)1}$$
$$\frac{30}{0} \quad \frac{14}{1} \quad \frac{6}{1} \quad \frac{2}{1} \quad \frac{0}{1}$$

Therefore, 30 is 11110 in binary. Then we change the fractional part to binary:

.75*2=1.50 ⇒ .50*2=1.00

Therefore, 30.75 in decimal is 11110.11 in binary.

IEEE (Institute of Electrical and Electronics Engineers)【电气和电子工程师学会】 FPS (Floating-Point Standard)【浮点标准】 has defined two standards for storing floating-point numbers in memory: single precision【单精度】 (32 bits), and double precision【双精度】 (64 bits), as shown in Table 3.3.

Table 3.3 IEEE standards for floating-point representation

Floating-point numbers' standards	Sign	Exponent	Mantissa
single precision (32 bits)	1 bit	8 bits	23 bits
double precision (64 bits)	1 bit	11 bits	52 bits

The IEEE standard defines 1 bit for sign, either 0 or 1. It defines floating point numbers in their normalized form. To normalize the mantissa, IEEE standard normalize the fixed-part of a real and use unsigned integer representation to store it. The number of digits used to shift the decimal point to left or right is indicated by the exponent, and this new representation is called excess representation【移码】. Excess representation is also known as biased representation. It adds a designated biased value to the original value to store all exponents as an unsigned integer【它在原值上，增加了一个指定的偏置值，以把所有指数，存储为一个无符号整数】. If the exponent occupies m bits in computer memory, the designated biased value is $2^{m-1}-1$ (referred to as L). The shifting in excess system with 4-bit allocation is shown in Figure 3.10. This new system is generally called Excess-L, like Excess-7.

For instance, we use IEEE 754 single precision format to represent −281.875. For single precision, the number is divided into sign bit, exponent, and fraction (also called significand or mantissa). The exponent is encoded as an 8-bit value, so the bias is 127 (or Excess-127). We proceed as follows:

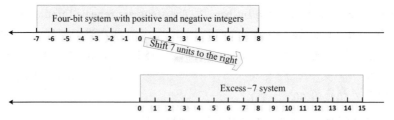

Figure 3.10 Shifting in Excess representation

(1) The sign is negative, so value of sign bit is 1, that is S = 1.

(2) Transform 281.875 to decimal: $(100011001.111)_2$.

(3) Normalization: $(100011001.111)_2 = (1.00011001111)_2 \times 2^8$.

(4) E is the exponent field and M is the mantissa. E = 8 + 127 = 135 = $(10000111)_2$ and M = $(000011001111)_2$. We need to add thirteen zeros at the right of M to make it 23 bits.

The final representation is shown in Figure 3.11.

Figure 3.11 −281.875 in IEEE 754 single precision format

Careful readers may notice that numbers with too small or too large absolute values cannot be stored with this representation. To handle this special case, it is agreed that the value equals 0 when the sign, exponent and mantissa are all 0 bits. Excess−127 gives an exponent range of 2^{-126} to 2^{+127}, and exponent 0 (2^{-127}) and 255 (2^{+128}) are reserved for special cases.

3.4 Data Storage

As previously described, numbers, text, images, audio and video are five different types of data. In Section 3.2 and Section 3.3, the methods of storing numbers inside a computer memory have been introduced. In this section, we will discuss the representation for other forms of data, normally called data encoding【数据编码】. In order to reduce the amount of storage space, the redundancy of data needs to be eliminated, and the technique used is called data compression【为了降低对存储空间的占用，要淘汰冗余的数据，所使用的技术就被称为数据压缩】.

3.4.1 Storing text

A text【文本】document is composed of various characters【字符】. In the English language, there are 26 letters. But considering the uppercase and lowercase, there are total 52 unique characters that need to be treated separately. Besides, 10 numeric characters, various punctuation characters, tab character, space character, and newline character, etc. also have to be represented【此外，还需表示 10 个数字字符、各种标点字符、制表符、空格字符、换行符等】. A character encoding is the mapping between the set of characters and the codes used to represent them. Two of the most widely used character sets【字符集】 ASCII (American Standard Code for Information Interchange)【美国信息交换标准码】 and Unicode 【统一码】 are introduced below.

1. ASCII and Unicode

ASCII uses 7-bits to represent 128 unique characters. ASCII reserves the first 32 codes for control characters, which are different from the following printable characters (x20 to x7E). It's easy to get a full ASCII table from the Internet【ASCII 码保留前 32 个编码作为控制字符，它们是与之后的可打印字符（X20 到 x7E）不同的。可以轻易地从互联网上获得一个完整的 ASCII 表】. But extended ASCII is not enough for other common languages, such as Chinese, which need symbols far beyond the 256 characters of the standard ASCII character set. To represent every character in the world for international use, the Unicode character set was developed. Unicode uses 16-bit (to represent over 65 thousand characters) and includes all the characters of the extended ASCII as its first 256 characters. Today, Unicode is widely used by many computer systems and programming languages【编程语言】.

2. Run-length encoding

Run-length encoding【行程长度编码】 is intended to compress the text with some characters repeating over and over again in a long sequence. The general idea is to replace a sequence of repeated characters with the repeated character followed by the repetition count. To indicate the start of the repeated characters, a flag character is needed. For example, a part of a DNA sequence is GGGGGAAAAAACCCC. Using run-length encoding with flag character '#', it can be represented by #G5#A6#C4. The number following the character specifies the number of times the character is appearing. The new compression sequence contains 9 characters instead of 15 in the original text.

3. Huffman encoding

Huffman encoding【哈夫曼编码】 is another text compression technique, based on the fact that some characters occur more frequently than others. The idea is to use shorter bit strings【字符串】 to replace the more frequent characters and leave longer bit strings to those appear less frequently. For simplicity, suppose a document only contains five kinds of characters: A, B, C, D, E, and the encoding table is listed in Table 3.4. Then the text AEBEAECADE can be encoded as 00 11 010 11 00 11 011 00 10 11. The original length for this text is 10*8 bits = 80 bits using ASCII code. After Huffman encoding, the length decreases to 22 bits, with a compression ratio of 0.275 (22/80).

Table 3.4 Encoding of characters

Character	Frequency	Huffman Code
A	16	00
B	10	010
C	15	011
D	29	10
E	30	11

3.4.2 Storing audio

We can divide the data types into two categories: analog data【模拟数据】 and digital data【数字数据】. Analog data is real world stuff like sounds, paintings, and temperature, which is not countable. Digital data, on the other hand, is discrete, and countable, like characters in text.

1. Digitization

Audio is a kind of analog data, while computers can only handle digital data, so audio data has to be sampled and then converted to numeric values. This process is sometimes called "digitize"【数字化】. For example, to represent the sound wave shown in Figure 3.12, a finite number of sample points on that curve are necessary. The number of sample points per second is called sampling rate【采样速率】. For each sample, quantization is the process that rounds the measured value to the closest integer value【对于每个采样，量化是将测量值四舍五入到最接近的整数值的过程】. Before being stored in the computer, each value should be converted to a bit pattern. The bit length for each sample value is sometimes called bit depth【比特深度】 or word length. The multiplication of sampling rate by the bit depth is normally called bit rate【比特率】.

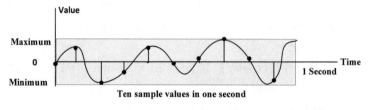

Figure 3.12　Digitization

2. Encoding format

There are several popular formats for audio encoding, like WAV (Wave)【波形声音】, MP3 (MPEG-2 Audio Layer 3)【MPEG第三层声音压缩技术】, MP4 (MPEG-4 Part 14), WMA (Windows Media Audio)【Windows 媒体音频】, Dolby AC-3 (Dolby Surround Audio Coding-3)【杜比AC-3环绕声】, AAC (Advanced Audio Coding)【高级音频编码】, DTS (Digital Theater Systems)【数字影院系统】, Dolby digital plus, and so on. Currently, MP3 is a dominant standard. This standard was designed by Motion Picture Experts Group (MPEG)【运动图像压缩标准】 which is used for compressing video data. If human could not hear the information, then it will be ignored by the compression method. There are also several standards, such as GSM (Global System for Mobile Communications)【全球移动通信系统】, G.729 and G.723.3. Besides, encoding standards for Android or iPhone like AMR (Adaptive Multi-Rate)【自适应多码率】 or iLBC (Internet Low Bitrate Codec)【互联网低比特率编解码器】 are widely used to transfer voice that many popular applications adopt.

3.4.3　Storing images

Techniques used by computers to store images are vector graphics【矢量图形】 and raster graphics【光栅图形】. Raster graphics is used to store photographs, which is an analog representation of an image. A photograph also needs to be digitized. Printed photographs taken with a digital camera may also be digitized. For audio, the intensity of data varies in time, but for images, it varies in space【对于音频来说，数据的强度随时间变化，但对于图像来说，数据的强度随空间变化】. Scanning is used as a sampling method for images. Its samples are normally called pixels【像素】 and the number of pixels used to represent the photo is called resolution【分辨率】. With high enough resolution, the discrete pixels can be continuous to human eye. As several encoding techniques can be used to handle the pixel's color, a pixel can be represented by different numbers of bits, which is called color depth【色深】. 24 bits are used by True-Color scheme to encode a single pixel, and JPEG (Joint Photographic Experts Group)【联合图像专家组】 uses this scheme to encode image and then compresses it to consume less bits. Indexed color scheme only uses a portion of these colors, and is used by GIF (Graphic Interchange Format)【图形交换格式】 to encode image. Main image formats are shown in Table 3.5.

Table 3.5 Main image formats

Image format	Brief description	Characteristics	Application scene	Extension name
BitMaP	BMP (BitMaP)【位图】is the standard image file format in Windows operating system, which uses bit map as storage format. In addition to optional image depth, it does not use any other compression	Support 1-24 color depth	Image software running on Windows	BMP
Personal Computer eXchange	PCX was developed by ZSOFT company in the developmen t of image processing software Paintbrush. It's a proprietary format for PC-based drawing program, and the general desktop publishing, graphic arts and video capture software supp ort this format	Run-length encoding	PC-based drawing programs	PCX
Tag Image File Format	TIFF was a generic image file format developed by Aldus and Microsoft for the desktop publishing system	Supports multiple en coding methods	Desktop publishing system, GI S, and remote sensing	TIFF
Tagged Graphics Analysis	TGA was developed by the company Truevision for its graphics card. It has been accepted by the international graphic image industry	Supports irregularly shaped graphics	The field of multi-media	TGA
Graphics Interchange Format	GIF was developed by CompuServe in 1987, its compression rate is generally about 50%, and almost all software support it	Can save multiple color images	The Internet, simple ani- mation	GIF
Joint Photographic Expert Group	JPEG is the network's most popular image format, developed by the Joint Photographic Experts Group. It is a lossy compression format, and can compress an image in a small storage space. JPG is short for JPEG, and jpg is a suffix, jpeg can be used as a suffix or to represent a file format	Variablecompression ratio	The Internet	JPEG
Exchangeable Image file Format	Promoted by company Fuji for digital camera in 1994, it is capable of storing photographic date, the use of aperture, flash exposure data and other inform ation	Stores exposure data, like photography date	Digital Cameras	EXIF
Kodak Flash PiX	FPX was jointly developed by Kodak, Microsoft, HP, and Live Picture, which has multi-resolution	With multiple resolution	Used by the Pict ure Easy Softwa re application included with Kodak digital cameras	FPX
Scalable Vector Graphics	It is based on XML (Extensible Markup Language)【可扩展标记语言】, developed by the World Wide Web Consortium's. And it can be arbitrarily enlarged while keeping very clear edge	Can enlarge graphics arbitrarily	Designing Web graphics pages of high resolution	SVG
Kodak PhotoCD	PCD is a Photo CD file format developed by Kodak. The format uses YCC color mode to define colors in the image	Uses YCC color mode	Save pictures on CD-ROM	PCD
Photoshop Document	PSD is a proprietary file format for an image processing software—Photoshop, and it can support a variety of image features such as layers, channels, masks and different color modes. It is a non-compressed format as it saves the original file	Retains all the original information	Image processing software named Photoshop	PSD

续表

Image format	Brief description	Characteristics	Application scene	Extension name
CorelDRAW	CDR is the dedicated graphics file format for a well-known graphics software CorelDRAW, which can record the file attributes, location and pagination, etc., but it is relatively poor in the degree of compatibility	Records properties, position and pagination of file	Mapping software named CorelDRAW	CDR
Drawing eXchange Format	DXF is a graphic file format of AutoCAD, which is stored in ASCII. It is very accurate on presenting graphics' size	Stores graphics in ASCII	When user needs to exchange CAD data between AutoCAD and other software	DXF
Ulead PhotoImapct	It is a dedicated image format for an image editing software named Ulead Photolmapct, and it is capable to record all the properties of image processed by Photolmapct	Replace layer with object to record information of image	Image editing software named Ulead Photoimpact	UFO
Encapsulated PostScript	It is a cross-platform standard format, mainly for storing vector and raster images	Using PostScript language to describe	Printing or printout	EPS
Adobe Illustrator	AI is a vector graphics file format for Adobe's software ILLUSTRATOR. It is a hierarchical file, each object in AI is independent and has its own property	Output in any size at the highest resolution	Vector software named Adobe illustrator	AI
Portable Network Graphics	The current version of PNG (Portable Network Graphics)【便携式网络图像格式】is the International Standard (ISO / IEC 15948: 2003), and published as a W3C recommendation in 2003. It can provide lossless compression	Supports 24-bit and 48-bit true color	JAVA program, web page and S60 program	PNG
High Dynamic Range Imaging	HDRI (High Dynamic Range Imaging) 【高动态范围成像】has larger brightness range than normal RGB format (only 8bit). It records brightness in the way of direct corresponding, which is different with the traditional image	Has the ability to save enough lighting information	When user need s environment li ghting informati on of the pictur e	HDRI
RAW Image Format	RAW file contains all the photo information of the original image file before it enters the camera's image processor after generated in the sensor. Many image processing software can process RAW file. The software provides adjustment of sharpness, white balance, gradation and color for a RAW format picture	Contains all photo information of the original image file	When user needs a fine picture	RAW

3.4.4　Storing video

Video is a sequence of images displayed one after another. Each image in video is called a frame 【帧】. Spatial compression【空间压缩】 and temporal compression【时间压缩】 are two types of compression techniques broadly used in video encoding. Spatial compression is done at the granularity【粒度】of frames. Each frame is independently compressed by removing redundant information. Temporal compression removes redundant frames directly, and spatial compression removes redundant data within each frame【每个帧，通过除去冗余的信息而被独立地压缩。时间压缩直接去除冗余帧，而空间压缩在每个帧内移除冗余数据】. In video compression, a frame is chosen as the key frame. Video compression only stores the key entirely and the changes of consecutive images between frames.

Video format is a recognition symbol that video player software imparts to a video file. Video format can be divided into two categories: local playing video and Internet streaming video. Although the stability and picture quality of the latter may not be as good as the former, Internet streaming video is widely used in video-on-demand【视频点播】, online presentations, distance education, and online video advertising, etc. Main video formats are shown in Table 3.6.

Table 3.6　Main video formats

Video format	Brief description	Video encoding	Audio encoding	Extension name	Popularity
Real Media	It was formulated by Realnetworks, and its compression ratio depends on network transmission rates	RealVideo9	RACC	rm/rmvb	Popular
Flash Video	It was formulated by Adobe Systems, and can be played by Adobe Flash Player version 6 and newer over the Internet	H.263	MP3	flv	Replaced by F4V
F4V	It was similar with Flash Video, but can support H.264 high-definition video encoding	H.264	MP3	f4v	Popular
Audio Video Interleave	AVI (Audio Video Interleave)【音频视频交叉存取格式】was released by Microsoft, which can be called conveniently with good image quality	MPEG-4	MP3	avi	Popular
Matroska	Matroska can integrate different types of audio tracks and subtitle tracks in a single file multiple, and its video encoding has very large degree of freedom	Many different types	Many different types	mkv	Popular
BHD	BHD (Black Hawk Down)【黑鹰坠落】is developed by Baofeng laboratory specifically for Baofeng player to play. It uses patented technology and is optimized for video quality, video size, video frame rate, bit rate, cellphone power saving, etc	FXV	FXA	bhd	Latest
MOV	MOV is the format of QuickTime movie for storing frequently used digital media type. It is developed by Apple	MPEG-4	MP3	mov	Classic
WMV	WMV (Windows Media Video)【Windows 媒体视频格式】is a technical standard developed by Microsoft for multimedia spreading through the Internet in real-time. It is an upgrade on the ASF (Advanced Stream Format) format. Its main advantages are: scalable media types, local or network playback, etc. The latest version of WMV is VC-1 standard	WMV	WMA	wmv	Classic

续表

Video format	Brief description	Video encoding	Audio encoding	Extension name	Popularity
MP4	MPEG (Moving Picture Experts Group，即MPE G)【动态图像专家组】-4 (ISO/IEC 14496) is an international standard based on the second generation of compression and encoding technology. It uses audio-visual media object as basic unit and adopts content-based compression encoding	MPEG-4	MP3	mp4	Popular
3GP	A multimedia standard developed by Third Generation Partnership Project (3GPP), so that users can use 3G mobile phones to enjoy high-quality audio, video, and other multimedia content	H.263	AMR_NB	3gp	Classic
WebM	WebM was proposed by Google, it is an open and free media file format. It is actually a new container format based on the development of Matroska. It includes VP8 video track and Ogg Vorbis audio track	VP8	Ogg Vorbis		Latest

Video encoding means converting a video file in a video format to another video format, through specific compression technologies. Video encoding standard is the basic standard of information field, which combines the excellent performance of various image encoding algorithms【视频编码标准是信息领域的基础标准，它结合了多种图像编码算法的优良性能】. Currently, the international video encoding standards is mainly divided into two series: MPEG series of standards developed by ISO (International Organization for Standardization)【国际标准化组织】 and IEC (International Electrotechnical Commission)【国际电工委员会】, and H.26x series of standards developed by ITU (International Telecommunication Union)【国际电信联盟】. In China, a proprietary video encoding standard – AVS (Audio Video coding Standard)【音视频编码标准】was developed leading by the Ministry of Industry and Information Technology. There are various standards for video encoding, as shown in Table 3.7.

Table 3.7　Standards for video encoding (Part A)

Standard	Brief description	Main technologies	Popularity
H.261	H.261 is mainly used in older video conferencing and video telephony products, which is developed by ITU-T in 1984. It is the first use of digital video compression standard	Inter-frame prediction based on motion compensation, 16x16 macro-block, discrete cosine transform on 8x8 sub-blocks	Outdated
H.263	In 1995, ITU-T launched H.263 for low bit rate video conferencing. Its encoding algorithm is basically the same with H.261, but added some improvements to improve encoding performance and error correction capability	2D prediction, motion compensation of half-pixel precision, syntax-based arithmetic encoding	Classic

续表

Standard	Brief description	Main technologies	Popularity
H.264	H.264 was proposed by Joint Video Team in 2003, it significantly improved compression ratio, and strengthened the treatment of errors and loss in IP network, mobile network	Two-layers encoding system, macro block division supporting unequal shape, multi-frame reference	Popular
HEVC	HEVC (High Efficiency Video Coding)【高性能视频编码】is a video compression standard developed by the JCT-VC organization, also called H.265. It can double the data compression ratio while keeping the same video quality, compared with MPEG-4. Using this standard, the resolution can reach 8192×4320	Redefined grammar for video image segmentation, intra-frame prediction supporting 33 directions, encoding unit of optional size	Latest
MPEG-1	MPEG-1 was developed by MPEG in 1992, used for encoding the active image and sound on digital storage medium like VCD, with digital rate of 1.5Mb/s	Bi-directional encoding scheme, pro gressive scanning image, motion vector of half-pixel precision	Outdated
MPEG-2	MPEG-2 was published in 1994, it is designed to get high image quality and higher transmission rates, providing transfer rate of 4Mbps to 100Mbps. It can be used for Digital Video Broadcasting (DVB), home DVD and high-definition television (HDTV)	Supporting interlace scanning video, four levels of encoded image resolution, five profiles based on various compression ratio	Classic
MPEG-4	Its first version was published in 1998 and second version in 1999. In addition to defining audio and video encoding standard, it also emphasizes the interactivity and flexibility of the multimedia system. It is mainly used in video telephony and television broadcasting. DivX, XviD and 3ivx video codec almost use the second part of the MPEG-4 technology	Video object based encoding, video object plane based encoding, video object layer based hierarchical encoding	Popular
AVS	AVS is a new generation of audio and video encoding standard developed by China. It doubles the compression efficiency when compared to MPEG-2, and can use a smaller bandwidth to transmit the same content	2D entropy encoding, pixel interpolation of 1/4 precision, loop filter	Latest
Microsoft RLE	It is developed by Microsoft for AVI format. It is an 8-bit encoding, which can only support up to 256 colors, using run length encoding compression algorithm	Run length encoding compression algorithm	Classic
Others	Sorenson 3 is a video codec used by Apple's QuickTime, Indeo Video was developed by Intel for video encoding, and Microsoft Video 1 was developed by Microsoft to compress analog video	Wavelet transform, forward predictive frame, overlap smoothing technique	Classic

High-resolution【高分辨率】 is closely related to the size of video file. The higher the resolution, the larger the file. Many video compression technologies have been invented, as shown in Table 3.7. High-resolution display device【显示器件】 is needed to display video files of high-definition【高清】. The resolution of a video file depends on the resolution of hardware device which creates the file, like camera【摄影机】, video camera【摄像机】, camcorder【录像机】, etc.

Display device has the property of display resolution (or screen resolution【屏幕分辨率】), which represents the precision of screen image【屏幕图像】, and refers to how many pixels【像素】 display device can display. Playback【播放】 of high resolution videos depend on the resolution supporting ability of display device【高分辨视频播放的实现，依赖于显示器件自身的分辨率的支持能力】. In the industry of display device, the manufacturers are used to adopt the horizontal resolution to define display resolution, because the device dimension ratio (that is, the ratio of horizontal scale to vertical scale)【显示比】 is fixed with 16:9 or 4:3. There are various display resolutions of an electronic image device, such as a computer monitor or television display, or image digitizing device, such as an image scanner, microscope, telescope, or camera. The industrial practices of display resolutions are shown in Table 3.8.

Table 3.8　Industrial practices【企业标准】 of display resolutions

Industrial practices	Brief introduction	Resolution
1K	When display device has horizontal resolution of approximately 1,000 (1 Kilo) pixels, it can be called 1K resolution	800×600, 960×600, 1024×768, 1280×800, 1440×900
2K	2K resolution generally refers to display content having horizontal resolution of approximately 2,000 (2 Kilo) pixels. Digital Cinema Initiatives (DCI) defined the standard of 2K resolution as 2048 × 1080	2048×1080, 1998×1080, 2048×858, 2560×1440, 2048×1536
3K	When display device has horizontal resolution of approximately 3,000 (3 Kilo) pixels, manufacturers usually call it 3K resolution	2880×1620, 3200×1800, 3072×1728
4K	4K resolution generally refers to display content having horizontal resolution of approximately 4,000 (4 Kilo) pixels and vertical resolution of approximately 2,000 pixels. Digital Cinema Initiatives (DCI) defined the standard of 4K resolution as 4096×2160	3840×2160, 4096×2048, 4096×2160, 4096×3112, 4096×3072

2K and 4K standards have been defined in movie projection industry. 4K resolution generally contains 4096×2160 pixels, which is four times of 2K projectors and High-Definition (HD)【高清】 television. It belongs to Ultra-High-Definition (UHD)【超高清】 resolution. At this resolution, the audience is able to see the screen in every detail, every close-up. Currently, TCL, Samsung, LG, Skyworth and other brands have launched 4K/UHD televisions.

3.5　Cloud Storage and Big Data Storage

3.5.1　Cloud storage

Cloud storage【云存储】 uses logical pools which use the physical storage of multiple servers in different locations to store the digital data【云存储，使用逻辑池来存储数字数据，而逻辑池使用多

个处于不同位置的服务器的物理存储】. As a new network storage technology, it integrates a large variety of different types of network storage devices to make them work collaboratively through a collection of software applications, using the cluster application, network technology, and distributed file systems. Thus cloud storage provides external data storage and an access service to users. For users, cloud storage is not a specific device, but rather a collection of storage devices and servers available as a service. Figure 3.13 shows a situation in cloud storage.

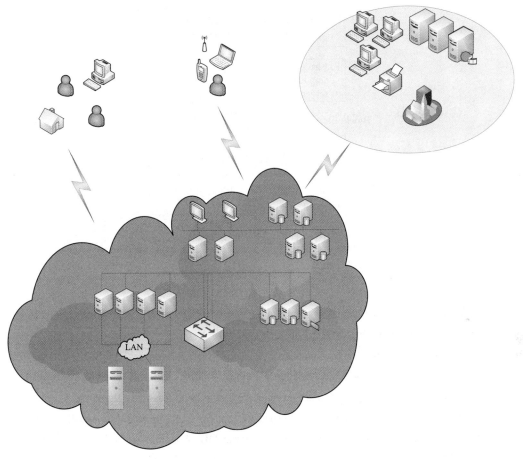

Figure 3.13　Cloud storage

Cloud storage usually refers to a hosted object【对象】 storage service, and other types of data storage that are available as a service, such as block storage. Object-based storage architecture manages data as objects on storage devices【云存储通常是指一个托管的对象存储服务，以及其他类型的可作为服务的数据存储，如块存储。基于对象的存储架构，是将数据作为存储设备上的对象来管理】 (Figure 3.14).

Amazon S3 (Simple Storage Service) and Microsoft Azure Storage are two typical instances of object storage services. OpenStack Swift is a typical instance of an object storage software. Numerous websites are providing cloud storage services now, and one of the most famous is Dropbox (www.dropbox.com). Before using Dropbox, an account needs to be created, and then the software needs to be installed on each of your computers, for taking advantage of the synchronization feature. Another popular Chinese cloud storage service is Tencent Weiyun, which offers a massive free cloud storage of 10 TB (10240 GB). Baidu Cloud, and 360 Cloud Drive are also popular Chinese cloud storage services that offers 2 TB, and 360 GB respectively of free storage. Western cloud storage services such as Dropbox, Google Drive, OneDrive, SFShare, etc. offers only 2 GB to 50 GB.

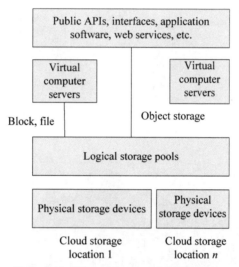

Figure 3.14   The architecture of cloud storage

For data model of cloud storage, there are two commonly used categories: relational and non-relational or NoSQL (Not Only SQL)【非关系型数据库】.

Cloud databases, such as Oracle Database, Microsoft SQL Server, and MySQL, can run on the cloud as a service or in a Virtual Machine Image【虚拟机镜像】. It is difficult for SQL databases to scale in a cloud environment, so the SQL database services are trying to address this challenge.

NoSQL data stores are data processing alternatives that can scale up and down easily. Examples of NoSQL data stores are Amazon's SimpleDB and Dynamo DB, Apache Cassandra, HBase, CouchDB, Couchbase, and MongoDB. NoSQL databases can support heavy read/write loads.

3.5.2   Big data storage

Big data【大数据】 is drawn from text, images, audio, video, and other storage-intensive multimedia formats. Big data size is constantly exploding, as of 2012 ranging from a few dozen terabytes to multiple petabytes of data (Figure 3.15).

There are three main storage technologies for big data.

1. Distributed file system

Distributed file system stores large-scale massive data in different storage nodes, in the form of files, and manages those files through a distributed system. In order to solve complex problems, this technology divides the large task into multiple smaller tasks, and uses multiple processors or multiple computer nodes in parallel computing to solve problems more efficiently【为了解决复杂的问题,这种技术将所述的大任务分成多个较小的任务,并且使用多个处理器或多个计算机节点进行并行计算,以便更有效地解决问题】. For example, the Hadoop framework uses HDFS (Hadoop Distributed File System)【Hadoop 分布式文件系统】which is scalable, distributed, and portable file system. Figure 3.16 shows big data storage with HDFS. Apache Spark is another cluster computing framework, which is based on a data structure called the resilient distributed dataset (RDD), and RDDs often be constructed by referencing data sets of external storage systems, like HDFS.

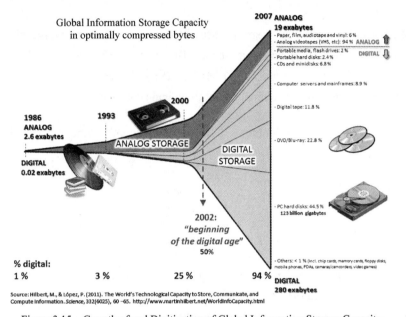

Figure 3.15 Growth of and Digitization of Global Information Storage Capacity

(Courtesy of http://www.martinhilbert.net/WorldInfoCapacity.html/)

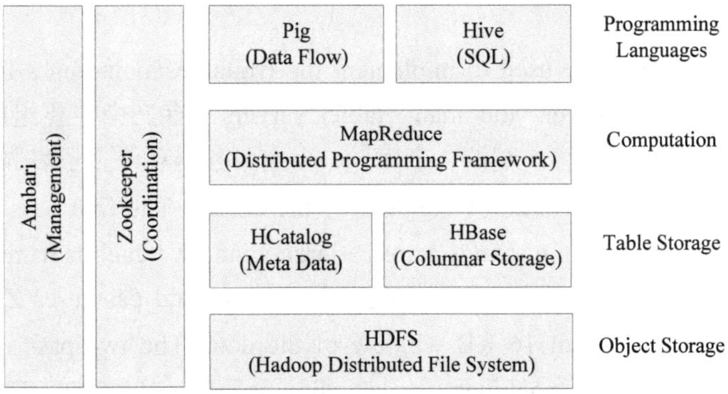

Figure 3.16 Hadoop uses HDFS for big data storage

2. Non-relational database

The distributed databases using the relational model cannot easily adapt to data storage in the era of big data, thus NoSQL databases are encouraged. NoSQL does not use SQL (Structured Query Language)【结构化查询语言】 as its query language and can store data without a predefined schema 【NoSQL 数据库,不使用 SQL 作为它的查询语言,它无需预定义的模式,就可存储数据】.

3. Data Warehouse

Data warehouse【数据仓库】commonly uses hardware and software in integrated manner, in order to provide the best performance. Such a database adopts new technologies which are more suitable for large data inquiry. Columnar storage and MPP (massively parallel processing)【大规模并行处理】 are two representative mature technologies.

3.5.3 Bigtable

1. Data Model

A Bigtable is a sorted map which is sparse, distributed, and persistent multidimensional. The map is indexed by a row key, a column key and a timestamp【BigTable 是一个稀疏的、分布式的、多维的有序映射。该映射由一个行键、一个列键和一个时间戳索引】. Bigtable treats each value in the map as an uninterpreted string. (row: string; column: string; time: int64) –> string

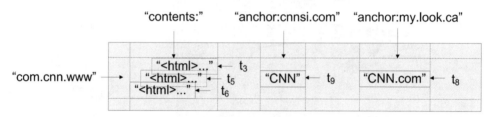

Figure 3.17   Bigtable map

A slice of an example table that stores Web pages is shown in Figure 3.17. The row name is a URL (Uniform Resource Locator)【统一资源定位符】 which is reverse to the original. The page contents are contained in the contents column family, and the anchors referring the web page are contained in the anchor column family【该网页的内容被包含在内容列族中，而引用该网页的锚，被包含在锚列族中】.

2. Compression

Three major components are used to implement the Bigtable, including a library that is linked into every client, a master server, and many tablet servers【有三个主要组成部分被用于实现Bigtable，包括被链接到每一个客户端的一个库、一个主服务器和许多子表服务器】. A two-pass custom compression scheme is commonly adopted by the clients. The first pass uses a compression scheme to compresses long common strings across a large window, which is named BMDiff (Bentley and Mcllroy's Scheme)【宾利和麦克罗伊压缩技术】. The second pass uses Zippy/Snappy, which searches for repetitions in a small 16 KB window of the data. The two-pass custom compression scheme is very fast. On modern machines, the encoding speed is 100-200 MB/s and the decoding speed is 400-1000 MB/s.

3.6   References and Recommended Readings

- Brookshear J G. Computer Science: An Overview. Pearson, 2015.

- Dale N, Lewis J. Computer Science Illuminated. Burlington, MA: Jones & Bartlett Learning, 2016.
- Forouzan B A, Mosharraf F. Foundations of Computer Science. Cengage Learning, 2008.
- Hadley J. When Online File Storage Gets Legal. Regulatory Compliance, 2014.
- Kurdi H, Li M, Al-Raweshidy H S. Taxonomy of Grid Systems. Grid and Cloud Computing: Concepts, Methodologies, Tools and Applications. IGI Global. 77-79, 2012.
- Miano J. Compressed Image File Formats: JPEG, PNG, GIF, XMB, BMP. Addison-Wesley Professional, 1999.
- Murray J D, vanRyper W. Encyclopedia of Graphics File Formats: The complete reference on CD_ROM with Links to Internet Resources. O'Reilly Media, 1996.
- Taubman D, Marcellin M. JPEG2000 Image Compression Fundamentals, Standards and Practice: Image Compression Fundamentals, Standards and Practice. The Springer International Series in Engineering and Computer Science, 2002.
- Chang F, Dean J, Ghemawat S, et al. Bigtable: A Distributed Storage System for Structured Data. ACM Transactions on Computer Systems (TOCS), 26(2): 4, 2008.
- Haghighat M, Zonouz S, Abdel-Mottaleb M. CloudID: Trustworthy Cloud-based and Cross-Enterprise Biometric Identification. Expert Systems with Applications, 42(21): 7905–7916, 2015.
- Jiang X, Wang W, Sun T,et al. Detection of Double Compression in MPEG-4 Videos Based on Markov Statistics. IEEE Signal Processing Letters, 20(5): 447-450, 2013.
- Sullivan G J, Ohm J R, Han W J ,et al. Overview of the High Efficiency Video Coding (HEVC) Standard. IEEE Transactions on Circuits and Systems for Video Technology, 22(12): 1649-1668, 2012.
- Vernik G, Shulman-Peleg A, Dippl S,et al. Data On-boarding in Federated Storage Clouds. Proceedings of the 2013 IEEE Sixth International Conference on Cloud Computing. IEEE Computer Society, 2013.
- Wiegand T, Sullivan G J, Bjontegaard G,et al. Overview of the H.264/AVC Video Coding Standard. IEEE Transactions on Circuits and Systems for Video Technology, 13: 560-576, 2003.
- ITU-T. Draft recommendation H.263 video coding for narrow telecommunication channels at below 64kbps. 1995.
- The Ultimate Guide to 4K Ultra HD, Ultra HDTV Magazine, 2013.
- Android Developer: Android Supported Media Formats, http://developer.android.com.
- AVS [S/OL]. http://www.avs.org.cn.
- IBM Knowledge Center, http://www.ibm.com/support/knowledgecenter.
- iOS Developer Library: Using Audio, https://developer.apple.com/library/ios.

3.7 Summary

- Numbers, text, audio, image and video are five types of data used in a computer. Computer

transforms the data into a uniform representation called a bit pattern.
- Integers represent numbers which have no fractional part. An unsigned integer can never be negative. Fixed-point representation is the method used for storing integers.
- Sign-and-magnitude format uses the leftmost bit (Most Significant Bit, also called sign bit) to represent the sign, and the rest of the bits define the magnitude.
- Two's complement is the most popular, the most important, and the most widely used way to represent signed integers today.
- A number which combines an integral part with a fractional part is called a real. To store real numbers in computer, the solution is to use floating-point representation, in which a number contains three sections: sign, exponent, and mantissa.
- A text document is composed of a sequence of characters. A bit pattern (ASCII and Unicode) can represent each character. In run-length encoding, the repeated character is replaced by one instance of that character followed by the number of its occurrences. In Huffman encoding, codes are assigned to characters such that the code length depends on the relative frequencies (or weight) at which each corresponding character occurred.
- Audio belongs to analog data which should be digitized before encoding. MP3 is the dominant standard for compressing audio.
- Techniques used by computers to store images are vector graphics and raster graphics (bitmap graphics). JPEG compression is the method of compressing pictures and graphics.
- Video is a series of consecutive images shown one after another. MPEG is one of the most widely used video compression method.
- Generally, 24 bits are used by True-Color scheme to encode a single pixel.
- BMP is the standard image file format in Windows operating system, which uses bit map as storage format. In addition to optional image depth, it does not use any other compression.
- GIF was developed by CompuServe in 1987, its compression rate is generally about 50%, and almost all software support it.
- JPEG is the network's most popular image format, developed by the Joint Photographic Experts Group. It is a lossy compression format, and can compress an image in a small storage space.
- BHD is developed by Baofeng laboratory specifically for Baofeng player to play. It uses patented technology and is optimized for video quality, video size, video frame rate, bit rate, cellphone power saving, etc. Audio Video Interleave (AVI) was released by Microsoft, which can be called conveniently with good image quality.
- Flash Video was formulated by Adobe Systems, and can be played by Adobe Flash Player version 6 and newer over the Internet. And F4V was similar with Flash Video, but can support H.264 high-definition video encoding.
- The international video encoding standards contain two main series: MPEG series of standards developed by ISO and IEC, and H.26x series of standards developed by ITU.
- HEVC is a video compression standard developed by the JCT-VC organization, also called H.265. It can double the data compression ratio while keeping the same video quality, compared with MPEG-4. Using this standard, the resolution can reach 8192×4320.

- Cloud storage uses logical pools which are based on the physical storage of multiple servers spanned across geographically separated locations.
- A Bigtable is a map which is sparse, distributed, and persistent multidimensional. Several clients use a two-pass custom compression scheme.
- Big data adopts three storage technologies, including distributed file system, non-relational database, and data warehouse.
- 4K resolution generally refers to display content having horizontal resolution of approximately 4,000 (4 Kilo) pixels and vertical resolution of approximately 2,000 pixels.

3.8 Practice Set

1. How a negative integer is represented using two's complement format?
2. How is a real number stored in floating point representation?
3. Represent the number $(24.076)_{10}$ in floating point binary representation. Assume the number is 32 bits long using 1 bit for the sign, 8 bits for the exponent, and 23 bits for the mantissa.
4. What is run-length encoding and Huffman encoding?
5. What is the data model of Bigtable?
6. Two bytes consists of _____ bits.
   a. 4   b. 8   c. 16   d. 32
7. In a set of 128 symbols, each symbol requires a bit pattern length of _____ bits.
   a. 5   b. 6   c. 7   d. 8
8. How many symbols can be represented by a bit pattern with eight bits? _____.
   a. 32   b. 64   c. 128   d. 256
9. If the leftmost bit in _____ number representation is 0, then the decimal number is positive.
   a. two's complement   b. floating point
   c. excess system   d. both a and b
10. For a fraction part, its exponential value can be stored by which method of number representation? _____.
    a. Unsigned integer   b. Two's complement
    c. Excess system   d. None of the above
11. When using excess representation, we _____ the bias number to the number to be converted.
    a. add   b. subtract
    c. multiply   d. divide
12. _____ defines the precision of the fractional part of a number stored in a computer.
    a. Exponent   b. Mantissa
    c. Sign   d. None of the above
13. A string of fifty 1s is replaced by two markers, a 1, and the number 50. This is _____.
    a. Run-length encoding   b. Morse coding

c. Huffman encoding d. Lempel Zip encoding

14. _____ commonly uses hardware and software in integrated manner, in order to provide the best performance.
    a. Distributed file system   b. Non-relational database
    c. Data Warehouse   d. SQL databases

15. Cloud storage stores the digital data in _____ pools.
    a. physical   b. logical   c. various   d. disk

16. AVS was developed by _____.
    a. China   b. ISO   c. IEC   d. ITU

17. H.26x series of standards for video encoding were developed by _____.
    a. China   b. ISO   c. IEC   d. ITU

18. _____ is a video compression standard developed by the JCT-VC organization, also called H.265.
    a. AVS   b. MPEG-2   c. HEVC   d. MPEG

19. _____ is the standard image file format in Windows operating system, which uses bit map as storage format. In addition to optional image depth, it does not use any other compression.
    a. GI   b. BMP   c. JPEG   d. PCX

20. Generally, _____ bits are used by True-Color scheme to encode a single pixel.
    a. 8   b. 12   c. 24   d. 48

21. Change the following decimal numbers to 16-bit unsigned integers. _____
    a. 52   b. 522   c. 2345   d. 532

22. Change the following decimal numbers to 8-bit two's complement integers. _____
    a. 82   b. −119   c. 53   d. 126

23. Change the following 8-bit unsigned numbers to decimal. _____
    a. 10010100   b. 01101011
    c. 11111001   d. 10101111

24. Change the following 8-bit two's complement numbers to decimal. _____
    a. 10001000   b. 00000011
    c. 10001011   d. 00110001

25. Use 32-bit IEEE format to represent the following numbers. _____
    a. 7.1875   b. −12.640625
    c. 11.40625   d. −0.375

26. The following numbers are sign-and-magnitude binary numbers in an 8-bit allocation. Convert them to decimal. _____
    a. 10001000   b. 00000011
    c. 10001011   d. 00110001

27. Use sign-and-magnitude with 8-bit allocation to represent the following decimal integers. _____
    a. 53   b. −107   c. −5   d. 154

28. 4K resolution generally refers to display content having horizontal resolution of approximately _____ pixels.

a. 1000 b. 2000 c. 3000 d. 4000

29. There is an alternative method to find the two's complement of a number. First, take the one's complement of the number. Then, add 1 to the result. Try both methods using the following numbers. Compare and contrast the results. _____

    a. 01110111   b. 11111100

    c. 01110100   d. 11001110

Unit 4
Data Operations

In this chapter, you'll learn how to operate on data. For different data types【数据类型】, there are corresponding operations【运算】 you can take. Actually, there are three types of data operations【数据运算】: logical operations【逻辑运算】, arithmetic operations【算术运算】, and shift operations【移位运算】. Our objective in this chapter is to learn and understand how to take these operations in different applications.

4.1 Logical Operations

In Chapter 3, we've learned that data inside a computer is stored as patterns【模式】 of bits【位】. The logic operations can be defined at the bit level【位层次】 and at the pattern level【模式层次】 and can be applied on data using Boolean operators【布尔运算符】: NOT, AND, OR, and XOR. The effect of applying these operators to each individual bit and to an n-bit pattern【n-位模式】 is the same. So the foundation of this section is logical operations at the bit level.

4.1.1 Logical operations at bit level

You can apply logical operations on a bit if you interpret this bit as a logical value. A logical value can only take one of two types of values: TRUE【真】 or FALSE【假】. In Boolean logic, 1 stands for TRUE, and 0 stands for FALSE. Logical operations can accept one or two binary bits as inputs producing one bit of the result. If an operator accepts only one input, then it is called unary operator【一元运算符】. If it works with exactly two inputs, then it is called binary operator【二元运算符】. There is only one unary operator and six binary operators, as shown in Figure 4.1.

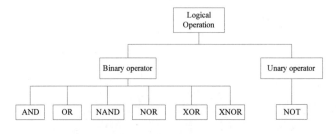

Figure 4.1 Logical operations

1. NOT operator

The NOT operator 【非运算符】 takes one input and produces one output, which is the complement of the input 【输入位取反】. Table 4.1 shows the truth table 【真值表】 for the NOT operation at bit level. A truth table lists all possible input combinations along with the corresponding outputs.

Table 4.1 Truth table for NOT operation

X	NOT X
0	1
1	0

2. AND operator

The AND operator 【与运算符】 accepts two inputs and creates one output. For each pair of input bits, the result is 1 if and only if both bits are 1【两个输入位均为 1 时，与运算结果才为 1】; otherwise, the result is 0. Table 4.2 shows the truth table for the AND operation at bit level.

Table 4.2 Truth table for AND operation

X	Y	X AND Y
0	0	0
0	1	0
1	0	0
1	1	1

Inherent Rule 【固有规则】 *of the AND Operator*: The AND operation can be used to force individual bits in a bit string to 0. From Table 4.2, we can see that if one of the input bits is 0, the result is always 0 regardless of the other input bit, which means you don't need to check the corresponding bit value of the other input 【如果一个输入位为 0，则与运算结果为 0，不必再检查另一个输入中相应的位】.

3. OR operator

The OR operator【或运算符】 accepts two inputs and creates one output. For each pair of input bits, the result is 0 if and only if both bits are 0【两个输入位均为 0 时，或运算的结果才为 0】; otherwise, the result is 1. Table 4.3 shows the truth table for the OR operation at bit level.

Table 4.3 Truth table for OR operation

X	Y	X OR Y
0	0	0
0	1	1
1	0	1
1	1	1

Inherent Rule of the OR Operator: The OR operation can be used to force individual bits in a bit string to 1. If one of the input bits is 1, the result is always 1 regardless of the other input bit, which means you don't need to check the corresponding bit value of the other input 【如果一个输入位为 1，或运算结果为 1，不必再检查另一个输入中相应的位】.

4. XOR operator

The XOR (exclusive OR) operator 【异或运算符】 accepts two inputs and creates one output. For each pair of input bits, the result is 0 if and only if both bits are equivalent【两个输入位相等时，异或运算的结果才为0】; otherwise, the result is 1. Table 4.4 shows the truth table for the XOR operation at bit level.

Table 4.4 Truth table for XOR operation

X	Y	X XOR Y
0	0	0
0	1	1
1	0	1
1	1	0

The XOR operator can be used to selectively invert bits in a bit string. If one of the input bits is 1, the result is always the inverse of the other input bit. If one of the input bits is 0, the result is exactly the value of the other input bit. Actually, the XOR operator is not a basic operator. We can simulate its functionality by using the three basic operators【使用三个基本运算符模拟异或运算符的运算功能】. The following expression shows the relation between them:

$$x \text{ XOR } y \Leftrightarrow [x \text{ AND } (\text{NOT } y)] \text{ OR } [(\text{NOT } x) \text{ AND } y]$$

5. XNOR operator

The XNOR operator【同或运算符】 accepts two inputs and creates one output. For each pair of input bits, the result is 0 if and only if both bits are not equal【当且仅当两个输入位不相等时，同或运算的结果才为0】; otherwise, the result is 1. Table 4.5 shows the truth table for the XNOR operation at bit level.

Table 4.5 Truth table for XNOR operator

X	Y	X XNOR Y
0	0	1
0	1	0
1	0	0
1	1	1

Like XOR operator, the XNOR operator can be represented by using the three basic operators. The following expression show the relation between them:

$$x \text{ XNOR } y \Leftrightarrow [x \text{ AND } y] \text{ OR } [(\text{NOT } x) \text{ AND } (\text{NOT } y)]$$
$$\Leftrightarrow \text{NOT } (x \text{ XOR } y)$$

6. NAND and NOR operator

The NAND【与非】 (or not-and) operator applies AND to its input bits and then inverts the output by applying NOT【与非运算符首先对输入位进行与运算，再应用非运算进行反转】. For each pair of input bits, the result of applying NAND operator is 0 if and only if both bits are 1; otherwise, the result is 1.

Similarly, the NOR【或非】 (or not-or) operator applies OR to its input bits and then inverts the output by applying NOT【或非运算符首先对输入位进行或运算，再通过非运算进行反转】. For each pair of input bits, the result of applying NOR operator is 0 if and only if either of its input bits are 1; otherwise, the result is 1.

They can be represented through following expressions:

$$x \text{ NAND } y \Leftrightarrow \text{NOT } [x \text{ AND } y]$$
$$x \text{ NOR } y \Leftrightarrow \text{NOT } [x \text{ OR } y]$$

The corresponding truth tables are shown in Table 4.6 and Table 4.7.

Table 4.6 Truth table for NAND operator

X	Y	X NAND Y
0	0	1
0	1	1
1	0	1
1	1	0

Table 4.7 Truth table for NOR operator

X	Y	X NOR Y
0	0	1
0	1	0
1	0	0
1	1	0

4.1.2 Logical operations at pattern level

The logical operators that can be applied to an *n*-bit pattern are the same as the logical operators applied to each individual input bit 【 *n*-位模式的逻辑运算就是把运算符应用到该模式中的每一位 】. There are some examples to help us understand this rule.

Example 4.1

Apply the NOT operator on the bit pattern 11101010.

Solution

```
       1 1 1 0 1 0 1 0   Input bits
  NOT ─────────────────
       0 0 0 1 0 1 0 1   Output bits
```

Example 4.2

Apply AND, OR, XOR, XNOR, NOR, and NAND operators on the bit patterns 10101101 and 01101011.

Solution

```
          1 0 1 0 1 1 0 1   Input1 bits                 1 0 1 0 1 1 0 1   Input1 bits
  AND     0 1 1 0 1 0 1 1   Input2 bits         OR      0 1 1 0 1 0 1 1   Input2 bits
          0 0 1 0 1 0 0 1   Output bits                 1 1 1 0 1 1 1 1   Output bits

          1 0 1 0 1 1 0 1   Input1 bits                 1 0 1 0 1 1 0 1   Input1 bits
  XOR     0 1 1 0 1 0 1 1   Input2 bits         XNOR    0 1 1 0 1 0 1 1   Input2 bits
          1 1 0 0 0 1 1 0   Output bits                 0 0 1 1 1 0 0 1   Output bits

          1 0 1 0 1 1 0 1   Input1 bits                 1 0 1 0 1 1 0 1   Input1 bits
  OR      0 1 1 0 1 0 1 1   Input2 bits         AND     0 1 1 0 1 0 1 1   Input2 bits
          1 1 1 0 1 1 1 1                               0 0 1 0 1 0 0 1
  NOT                                           NOT
  NOR     0 0 0 1 0 0 0 0   Output bits         NAND    1 1 0 1 0 1 1 0   Output bits
```

4.1.3 Logic Diagram Symbol

Logic operators include the corresponding LDS (Logic Diagram Symbol)【逻辑图符号】 and Logic Function Expression 【逻辑函数表达式】. IEEE standard 【电气电子工程师协会标准】 and IEC standard 【国际电工委员会标准】are two different universal standards for LDS. In the IEC standard, the symbol '1' means that the one input must be active【符号'1'表示该输入必须是激活的】. The '&' represents AND function. The '≥1' indicates that at least one active input is needed to activate the output【'≥1' 表示至少需要一个激活的输入才能激活输出】. The '=1' denotes that one and only one input must be active to activate the output. And the last symbol '=' implies that all inputs must stand at the same state【最后一个符号'='表示所有的输入必须处于相同的状态】. The LDS and Logic Function Expression are widely used in Digital Electronics【数字电子电路】and Integrated Circuit【集成电路】. Table 4.8 shows the details.

Table 4.8 Logic Diagram Symbol and Logic Function Expression

Logic Operator	ANSI and IEEE standard	IEC standard	Logic Function Expression
NOT			$\overline{A}$
AND			$A \cdot B$
OR			$A+B$
NAND			$\overline{A \cdot B}$
NOR			$\overline{A+B}$
XOR			$A \oplus B$
XNOR			$\overline{\overline{A} \cdot B}$ or $A \odot B$

4.1.4 Applications

We can modify a bit pattern with four binary logic operators: NOT, AND, OR, and XOR. These binary operators can complement, unset【复位】, set【置位】, or flip【反转】 the target bit pattern with a mask【掩码】. The mask is used to modify the target bit pattern.

1. AND: Unsetting specific bits

The most important application of AND operator is to unset (force to 0 【置0】) specific bits in a target bit pattern. To do that, we must use an unsetting mask【复位掩码】of the same length as of the target bit pattern. In this technique, to unset a specific bit of the target bit pattern, the corresponding bit in the mask is set to 0【对于目标位模式中需要置0的位，掩码的相应位设为0】. Otherwise, to leave a bit in the target bit pattern unchanged, the corresponding bit in the mask is set to 1. For example, if you want to unset the 4 rightmost bits of an 8-bit pattern, the unsetting mask must be 11110000 (Figure 4.2).

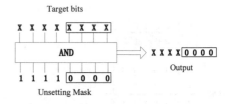

Figure 4.2 Example of unsetting specific bits of a bit pattern

2. OR: Setting specific bits

Similarly, we use OR operator to set (force to 1【置1】) specific bits in a target bit pattern. In this technique, to set a specific bit of the target bit pattern, the corresponding bit in the setting mask【置位掩码】 is set to 1【对于目标位模式中需要置1的位，掩码的相应位设为1】. Otherwise, to leave a bit in the target bit pattern unchanged, the corresponding bit in the mask is set to 0. An example is shown in Figure 4.3.

3. XOR: Flipping specific bits

The XOR operator is used to flip specific bits (0 becomes 1 and vice versa). The flip operation is functionally similar to the NOT operator. To flip a specific bit in the target bit pattern, the corresponding bit in the flipping mask【反转掩码】 is set to 1【对于目标位模式中的需要反转的位，掩码的相应位设为1】. Otherwise, to leave a bit in the target bit pattern unchanged, the corresponding bit in the mask is set to 0. Figure 4.4 shows an example of flipping specific bits.

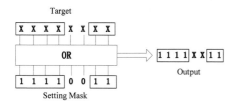

Figure 4.3　Example of setting specific bits of a bit pattern

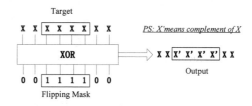

Figure 4.4　Example of flipping specific bits

There are interesting applications of XOR operator. XOR can be used to exchange two numbers without using any intermediate variable【异或运算能够在不引入中间变量的情况下交换两个数字的值】. If we have a=10011001 and b=10110101, then after taking three XOR operations, we will finally exchange the value of a and b.

$$a \Leftarrow a\ XOR\ b = 10011001\ XOR\ 10110101 = 00101100,$$
$$b \Leftarrow a\ XOR\ b = 00101100\ XOR\ 10110101 = 10011001,$$
$$a \Leftarrow a\ XOR\ b = 00101100\ XOR\ 10011001 = 10110101.$$

Other interesting applications of XOR operator are to do parity check of bit strings and to verify whether bytes sent across a network were corrupted and need retransmission, or they were sent without errors【异或运算的另一个有趣应用是进行位串的奇偶校验以验证通过网络发送的字节是发生了错误而需要重发还是发送无误】.

4.2　Arithmetic Operations

The four basic arithmetic operations are addition【加法】, subtraction【减法】, multiplication【乘法】, and division【除法】. These operations can apply to both integers【整数】 and floating-point numbers【浮点数】.

4.2.1　Arithmetic operations on integers

Among the four basic arithmetic operations, the addition and subtraction operations are fundamental and essential. So we focus on addition and subtraction operation of integers in this section. The corresponding procedure of multiplication and division operations can be found in books on computer architecture【计算机体系结构】, such as "Computer Organization and Architecture: Designing for performance".

1. Addition and subtraction in two's complement

All modern computers represent integers in the two's complement【二进制补码】 form. Addition and subtraction in two's complement are more or less the same as in decimal【十进制】. To get the two's complement negative notation of an integer, or binary representation of a negative number, write out the number in binary. Then invert the digits, and add one to get the result.

When the result of binary addition of any two bits is two (1 plus 1 is $(10)_2$), the resultant bit is 0 and we propagate a 1 to the next most significant place. This 1 is called carry【进位】. Now we can define the rules for adding two integers in two's complement form: adding 2 bits and propagating the carry to the next column from right to left【两个位相加，如果有进位，就加到左边的下一列上】. If there is a final carry after the leftmost column addition, we can discard【丢弃】 it.

It is important to note that every time the computer encounters the subtraction operation, it carries out addition operation instead of subtraction operation based on the following equation:

$$A - B = A + (-B) = A + (\overline{B} + 1)$$

Where $\overline{B}$ means the one's complement【二进制反码/一的补码】 of $(-B)$, and $(\overline{B}+1)$ means the two's complement of $(-B)$.

Example 4.3

Add two numbers in two's complement format: $(+35) + (+29) \rightarrow (+64)$

Solution

$(+35)$ and $(+29)$ in two's complement form are represented as 00100011 and 00011101 for an 8 bit memory location. The result remains same for any location size.

```
Carry        1 1 1 1 1
             0 0 1 0 0 0 1 1    +35
         +   0 0 0 1 1 1 0 1    +29
Result       0 1 0 0 0 0 0 0    +64
```

The result is 64 in decimal.

Example 4.4

Suppose $A = (-30)$ and $B = (+34)$. Show how B is subtracted from A, that is $A - B$.

$A = (11100010)_2 \quad B = (00100010)_2$

Solution

$A - B = -30 - 34 = (-30) + (-34)$.

$B = (34)_{10} = (00100010)_2$

To get the negative of B, we take its one's complement.

$\overline{B} = (11011101)_2$

Now, add 1 to get the two's complement.

$\overline{B} + 1 = (11011110)_2$

Using the same steps as above, we can get the binary of A which is a negative integer.
$A = (11100010)_2$
According to this example, $A - B = A + (\bar{B} + 1)$

```
Carry     1 1 1 1 1 1
Discard   1 1 1 0 0 0 1 0    A
       +  1 1 0 1 1 1 1 0    -B
Result    1 1 0 0 0 0 0 0
```

In the result, the leftmost bit 1 indicates that the number is negative, so take its two's complement to get $R = (11000000)_2$, which is −64 in decimal.

Example 4.5

Integer $A = (+127)$ and $B = (+3)$ are stored in two's complement form. Show how B is added to A.
$A = (01111111)_2$, $B = (00000011)_2$

Solution

A is added to B and the result is stored in R.

```
Carry     1 1 1 1 1 1
          0 1 1 1 1 1 1 1    A
       +  0 0 0 0 0 0 1 1    B
Result    1 0 0 0 0 0 1 0    R
```

Obviously, we get an error here: $A + B = R = -126 \neq 130$. We will discuss this issue in next section.

2. Overflow

When you try to store a value that is out of the range of the allocation, an arithmetic overflow【上溢】 occurs【当试图把一个数存储在超出地址分配所定义的范围时会发生算术上溢】. It is similar to what happened when you try to pour milk into a cup more than its capacity. The computer offers us a variety of "cups" with limited and fixed "volume" for storing "milk", which is number here.

There are two kinds of overflow: positive overflow【正上溢】 and negative overflow【负上溢】.When we represent an integer x, in two's complement form using N bits, the range of values that can be represented is $-2^{N-1} \leqslant x \leqslant 2^{N-1} - 1$. A negative overflow occurs when a number is less than -2^{N-1}, and a positive overflow happens when a number is greater than $2^{N-1} - 1$. Figure 4.5 shows these two types of overflows.

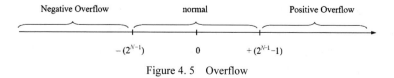

Figure 4.5 Overflow

Back to Example 4.5, the normal range is -2^{8-1} to $2^{8-1}-1$, which is -128 to 127. But the result (130) of the addition operation is out of this range. Then positive overflow occurs. The reason that actual result is -126 but not 130 is because of the natural feature of two's complement representation【因为二进制补码表示的固有特性，所以实际的结果是−126 而不是 130】.

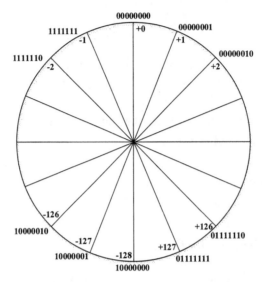

Figure 4.6　Two's Complement representation (clockwise)

(Courtesy of http://baike.baidu.com/view/377340.htm)

As we can see from Figure 4.6, when we reach the number 127 and add 3 to 127, the final result is -126 rather than 130.

4.2.2　Arithmetic operations on reals

All arithmetic operations can be applied to reals【实数】 stored in floating-point format. In this section, we only show rules for addition and subtraction of reals.

1. Addition and subtraction of reals

Addition and subtraction operation rules for floating-point numbers are basically the same. The basic steps are as follows:

- Check the signs【符号】.

  a) If the signs are the same, add the numbers and assign the common sign to the result【如果符号相同，相加其值，结果符号与它们相同】.

  b) If the signs are different, take the absolute values【绝对值】 of the numbers, subtract the smaller from the larger, and use the sign of the larger to give the result【如果符号不同，比较绝对值，绝对值大的减去小的，结果符号取绝对值大的一方】.

- When converting to scientific notation, move the decimal points【小数点】 to make sure the exponents【指数】 are the same. This means if the exponents are not equal, the decimal point of the number with the smaller exponent is shifted to the left to make both numbers having the same exponent【也就是说，当指数不同时，数值小的一方将小数点左移，使指数相同】.

- Add or subtract the mantissas 【尾数】 (including the whole part 【整数部分】 and the fractional part【小数部分】).
- Normalize【规范化】 the result before storing the result in memory【内存】.
- Check for any overflow and underflow.

Example 4.6

Show how the computer finds the result of (+21.75) + (+16. 4375) = + 38.1875.

Solution

$A = (+21.75)$, $B = (+16.4375)$. They are stored in floating-point form.

	S	E	M
A	0	1 0 0 0 0 0 1 1	01011100000000000000000
B	0	1 0 0 0 0 0 1 1	00000111000000000000000

After de-normalization【去规范化】, we have:

	S	E	De-normalized M
A	0	1 0 0 0 0 1 0 0	10101110000000000000000
B	0	1 0 0 0 0 1 0 0	10000011100000000000000

Because the exponents are the same, so we don't need to align【对齐】. We apply addition operation on the combinations of sign and mantissa. Then we have following result:

	S	E	De-normalized M
R	0	1 0 0 0 0 1 0 1	10011000110000000000000

Now the result need to be normalized. We decrement the exponent and shift the de-normalized mantissa one position to the left:

	S	E	De-normalized M
R	0	1 0 0 0 0 1 0 0	00110001100000000000000

So the final result is $R = +2^{132-127} \times 1.001100011 = + 38.1875$, as expected.

2. Underflow

In this section, underflow 【下溢】 refers to arithmetic underflow 【算术下溢】 or floating point underflow【浮点数下溢】, which means result of an arithmetic operation of floating-point number is less than the computer can actually store in memory【这表示浮点数算术运算结果小于计算机内存实际能够存储的数】. While, overflow means that an operation result is out of the normal range that an N-bit pattern can represent, not a computer system.

In Figure 4.7, the fminN means the minimum positive number that a computer system can represent, which is a number very close to zero point 【零点】. Then, the interval 【区间】 (−fminN, fminN), is called underflow gap【下溢间距】. In early design of computer system, any number inside the underflow gap would be set to zero. That kind of process is called "flush-to-zero underflow"【归零】. We usually use natural logarithm【自然对数函数】 to avoid arithmetic underflow issues.

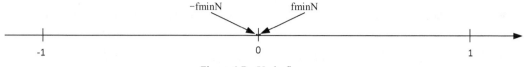

Figure 4.7　Underflow gap

4.3　Shift Operations

The shift operation is another common operation on bit level. There are two types of shift operations: logical shift operations【逻辑移位运算】, and arithmetic shift operations【算术移位运算】.

4.3.1　Logical shift operations

In the binary representation【二进制表示】 of signed number【有符号数】, the leftmost bit represents the sign of the number. But shift operation may change that bit【移位运算可能改变该符号位】. So a logical shift operation is usually applied to unsigned number【无符号数】. As described below, we distinguish two kinds of logical shift operations: logical shift, and logical circular shift.

1. Logical shift

In the right-shift logical operation【逻辑右移运算】, we discard the rightmost bit at first, then shift every bit to the right, and insert a 0 bit to the leftmost bit finally【逻辑右移运算中，丢弃最右位，然后将其他位向右移动一个位置，最左位填0】. The left-shift logical operation【逻辑左移运算】 works conversely. Figure 4.8 shows these two operations on 8-bit patterns.

2. Logical circular shift

The circular right-shift operation【循环右移运算】 shifts every bit to the right. The leftmost bit is replaced by the rightmost bit. The circular left-shift operation【循环左移运算】 works conversely. Figure 4.9 shows these two operations on 8-bit patterns.

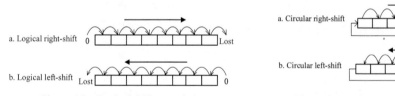

Figure 4.8　Logical shift operations　　　　Figure 4.9　Circular shift operations

Example 4.7

Take a logical left-shift operation and circular left-shift operation on the bit pattern 10111100 respectively.

Solution

For the logical left-shift operation, the leftmost bit is discarded and a 0 is inserted as the rightmost bit.

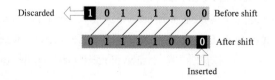

For the circular left-shift operation, the rightmost bit is replaced by the leftmost bit.

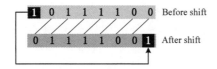

Example 4.8

Use a combination of logical and shift operations to find the value (0 or 1) of the third bit (from left to right) in a pattern of eight bits【利用逻辑运算和移位运算找出 8-位模式中的从左往右数第三个位】.

Solution

Use the mask 00100000 to AND with the target bits to access the third bit and set the rest of the bits to zero.

```
Target    a b c d e f g h
Mask      0 0 1 0 0 0 0 0    AND
Result    0 0 c 0 0 0 0 0
```

To get the value of the third bit (c), we can shift the new pattern five times to the right:

$$0\,0\,c\,0\,0\,0\,0\,0 \to 0\,0\,0\,c\,0\,0\,0\,0 \to 0\,0\,0\,0\,c\,0\,0\,0 \to 0\,0\,0\,0\,0\,c\,0\,0$$
$$\to 0\,0\,0\,0\,0\,0\,c\,0 \to 0\,0\,0\,0\,0\,0\,0\,c$$

If the result is 1, then the original third bit is 1. If the result is 0, then the original third bit is 0.

4.3.2　Arithmetic shift operations

Unlike the logic shift operations, arithmetic shift operations focus on signed integer【带符号位整数】 in two's complement form【与逻辑移位运算不同，算术移位运算用于二进制补码形式表示的带符号位整数】. Functionally, arithmetic right-shift operation【算术右移运算】 divides an integer by two, while arithmetic left-shift operation【算术左移运算】 multiplies an integer by two【从功能上来看，算术右移被用来对整数除以 2，而算术左移被用来对整数乘以 2】.

The arithmetic right-shift operation shifts every bit to the right. The leftmost bit remains unchanged and the rightmost bit is lost【最左边的符号位保持不变，丢弃最右位】. The arithmetic left-shift operation shifts every bit to the left. The Most Significant Bit (MSB)【最高有效位】 is lost and a 0 is inserted as the rightmost bit. Figure 4.10 shows the arithmetic shift operations on 8-bit patterns.

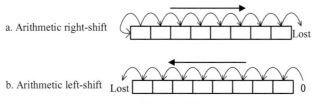

Figure 4.10　Arithmetic shift operations

Example 4.9

Use an arithmetic right-shift operation on the bit pattern 10110000. The pattern is an integer in two's complement form.

Solution

Note that the MSB remains unchanged.

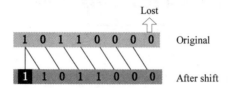

We can see that the result is -40, which is exactly the half value of original value, −80.

Example 4.10

Use arithmetic left-shift operation on the bit pattern 11010000. The pattern is an integer in two's complement form.

Solution

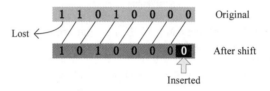

The original number is −48 and the new number is −96. Obviously, the original number is multiplied by two.

4.4 References and Recommended Readings

For more details about the subject discussed in this chapter, the following books are recommended.
- Dale N , Lewis J. Computer Science Illuminated. Jones & Bartlett Learning, 2014.
- Patterson D A , Hennessy J L. Computer Organization and Design: The Hardware/Software Interface. Morgan Kaufmann Publishers, 1998.
- Bryant R E , O'Hallaron D. Computer Systems: A Programmer's Perspective. Prentice Hall, 2008.
- Stalling W. Computer Organization and Architecture, Upper Saddle River. NJ: Prentice Hall, 2000.
- Floyd T L. Digital Fundamentals. Prentice Hall, 2008.
- https://en.wikipedia.org/wiki/Logic_gate.

4.5 Summary

- Integers are represented in the two's complement format in most computers.
- The carry is discarded if it is obtained after addition of the leftmost bits.

- Just negate the number to be subtracted and added when we execute subtraction operation in two's complement.
- Overflow occurs when a number is not within the range defined by the bit allocation.
- Logical operations includes unary operations (that take one input) and binary operations (that take two inputs).
- The result of the binary AND operator is true if both inputs are true. The result of the binary OR operator is false if both inputs are false. The result of the binary XOR operator is false if both inputs are the same.
- A mask is a bit pattern that is applied to a target bit pattern to get a particular result.
- Set the corresponding mask bit to 0 and use the AND operator if you need to unset (clear) a bit in a target bit pattern.
- Set the corresponding mask bit to 1 and use the OR operator if you want to set a bit in a target bit pattern.
- Set the corresponding mask bit to 1 (0) and use the XOR (XNOR) operator if you wish to flip a bit in a target bit pattern. Besides, you can exchange two number with three XOR operations, without using another intermediate variables.
- Logic operators including NAND, NOR, XOR, and XNOR are logic combination of the three basic logic operators, including NOT, AND, and OR.
- Logic operators have the corresponding Logic Diagram Symbol and Logic Function Expression in computer science field. It's widely applied in digital electronics.
- Overflow and underflow are two common errors that usually occur in arithmetic operations. But they are totally different issues. Underflow means that the operation results of floating-point number is less than the minimum value that a computer system can represent. While overflow means that an operation result is out of the normal range that an N-bit pattern can represent, not a computer system.
- Logical shift operators are usually applied to unsigned numbers while arithmetic shift operators are mainly applied to signed numbers. Arithmetic shift operations are often used to multiply or divide by constant powers of 2.

4.6 Practice Set

1. What happens to a carry from the most significant bit column in the addition of integers in two's complement notation?
2. What is overflow and why it occurs? What are the types and consequences of overflow?
3. What is the difference between overflow and underflow? What a computer system usually do when an underflow occur?
4. How do you adjust the representation of numbers with different exponents in the addition of floating-point numbers?
5. What binary operations can be used to set, unset, and flip bits? What bit patterns should the

masks have?

6. _____ are the connection expressions between the following logic operator groups.
   a. XOR, AND, OR, NOT
   b. NAND, AND, NOT
   c. XNOR, XOR
   d. XNOR, AND, OR, NOT

7. Draw the corresponding logic diagram symbol for the following logic operators. _____.
   a. NOT
   b. AND
   c. OR
   d. XOR
   e. NAND

8. Write down the right logic operator for following logic function expressions. _____.
   a. $A \odot B$
   b. $A + B$
   c. $\overline{A \cdot B}$
   d. $A \oplus B$
   e. $\overline{A}$

9. The _____ method of integer representation is the most common method to store integers in computer memory.
   a. sign-and-magnitude【原码】
   b. one's complement
   c. two's complement
   d. signed integers

10. In two's complement addition, if there is a final carry after the leftmost column addition, _____.
    a. add it to the rightmost column
    b. add it to the previous column
    c. discard it
    d. replace the current column

11. For an 8-bit allocation, the smallest decimal number that can be represented in two's complement form is _____.
    a. −512
    b. −126
    c. −127
    d. −128

12. In two's complement representation with a 4-bit allocation, you get _____ when you add 1 to 7.
    a. 8
    b. −9
    c. −8
    d. −0

13. If the exponent in Excess−127 is binary 10000101, the exponent in decimal is _____.
    a. 6
    b. 7
    c. 8
    d. 9

14. If you are adding two numbers, one of which has an exponent value of 4 and the other has an exponent value of 6, you need to shift the decimal point of the smaller number _____.
    a. one place to the left
    b. one place to the right
    c. two places to the left
    d. two places to the right

15. For the binary _____ operator, if both the inputs are 1, the output is 0.
    a. AND
    b. OR
    c. XOR
    d. XNOR

16. You use a bit pattern called a _____ to modify another bit pattern.
    a. mark
    b. carry
    c. float
    d. mask

17. To flip all the bits of a bit pattern, make a mask of all 1s and then _____ the bit pattern and the mask.
    a. AND
    b. OR
    c. XOR
    d. NOR

18. To set (force to 1) all the bits of a bit pattern, make a mask of all 1s and then _____ the bit pattern and the mask.
    a. AND
    b. OR
    c. XOR
    d. NOT

19. Using an 8-bit allocation, first convert each of the following numbers to two's complement, do the operation, and then convert the result to decimal. _____.
    a. 60 + 35   b. 60 − 35   c. −60 + 35   d. −60 −35

20. Which of the following operations creates overflow if the numbers and the result are represented in 8-bit two's complement notation. _____.
    a. 11000010 + 00111111   b. 00000010 + 00111111
    c. 11000010 + 11111111   d. 00000010 + 11111111

21. Show the result of the following floating-point operations. First convert each number to binary notation and do the operation; then convert the result back to decimal notation. _____.
    a. 34.075 + 23.12   b. −12.067 + 451.00
    c. 33.65 − 0.00056   d. −344.23 − 123.8902

22. In which of the following situations does an overflow never occur? Justify your answer. _____.
    a. Adding two positive integers
    b. Adding one positive integer to a negative integer
    c. Subtracting one positive integer from a negative integer
    d. Subtracting two negative integers

23. Show the result of the following operations:
    a. NOT (x9E OR x9E)
    b. xCC OR (NOT x00)
    c. (x99 AND x55) OR (x00 AND xFF)
    d. (x91 OR x33) AND (x00 OR xFF)
    e. (x3E XOR xF0) AND (x27 NOR xD2)
    f. (xAA XNOR x93) XOR (x35 NAND x10)

24. You need to flip the 5 rightmost and the 3 leftmost bits of a pattern of 16 bits. Show the mask and the operation.

25. You need to unset the 3 leftmost bits and set the 2 rightmost bits of a pattern. Show the masks and the operation.

26. Extract the fourth and fifth bits of an unsigned number in an 8-bit allocation by using a combination of logical and shift operators.

Unit 5
Computer Components

In this chapter, we describe the primary components that make up a stand-alone computer and the connections among them. The primary components are divided into three parts: the processing unit 【处理单元(器)】, main memory 【主存储器】 and the input/output subsystem 【输入/输出子系统】. In addition, we introduce machine cycle【机器周期】 and computer architecture 【计算机架构】.

5.1 Three Main Components

We will describe three main components that makes up a computer: the processing unit, the main memory and the input/output subsystem. Figure 5.1 shows the three broad categories of a standalone computer.

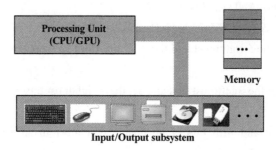

Figure 5.1   Computer hardware (subsystems)

5.1.1 Processing unit

1. Uniprocessor and multi-core processor

Processor【处理器】 is a component that performs operations on program instructions【指令】 and the most common processor is CPU (Central Processing Unit) 【中央处理单元(器)】. The processor consists of uniprocessor 【单处理器】 and multi-core processor 【多核处理器】.

Uniprocessor has only one processor per system. However, multi-core processor refers to an integrated circuit (IC)【集成电路】to which two or more complete and independent processor cores【核】 are attached. For example, AMD Phenom II X2 and Intel Core Duo are two-core (dual-core) processors,

Intel's i5 and i7 processors have four cores (quad-core). Multi-core processors are widely applied in general-purpose, embedded【嵌入式】, network, digital signal processing (DSP)【数字信号处理】, graphics processing unit (GPU)【图形处理单元（器）】, mobile phones, laptop, and supercomputer【超级计算机】.

It is known to us that China's new supercomputing system, Sunway TaihuLight, won the top of 47th TOP 500 list of the world's top supercomputers on June 20, 2016. With floating point speed of 93 petaflops per second (or 93 quadrillion floating point operations per second), Sunway TaihuLight is twice as fast and three times as energy-efficient as China's Tianhe-2【2016年6月20日，中国的"神威·太湖之光"荣获了第47届世界超算大会第一名，它的浮点运算速度为每秒9.3亿亿次，其速度是天河二号的2倍，效率也提高了3倍】. Tianhe-2 at 33.9 petaflop/s, the world's fastest supercomputer for three consecutive years (in six consecutive rankings), is now in the second position. The Sunway TaihuLight uses Sunway SW26010 chip designed and made in China【神威·太湖之光采用中国自行设计和制造的神威 SW26010（申威26010）芯片】. The CPU vendor【供应商】 is the Shanghai High Performance IC Design Center. Sunway TaihuLight system has 10,649,600 computing cores comprising 40,960 nodes and has 1.31 PB (or 32 GB) of primary memory for each node. Each node in Sunway TaihuLight has one SW26010 processor (or chip) that produces speeds of 3.06 teraflops/s with 260 cores and uses ShenWei-64 instruction set【神威·太湖之光每一个节点是一个申威 SW26010 处理器，该处理器有260个处理核，单片峰值性能为3.06Tflops/s(每秒3.06万亿次浮点运算次数)，并且使用了申威64指令集】. So Sunway SW26010 is a 1.45 GHz many-core 64-bit RISC processor. Figure 5.2 (a) shows Sunway TaihuLight System and Figure 5.2 (b) shows Sunway SW26010 processor that makes up the node.

(a) Sunway TaihuLight supercomputer (b) Sunway SW26010 many-core processor

Figure 5.2　Sunway TaihuLight supercomputer and SW26010 processor

Multi-core processors can run multiple threads simultaneously【多核处理器同时运行多个线程】. The throughput【吞吐量】, energy efficiency【能源效率】, and multitasking【多任务处理】 performance of multi-core processors depends on the proper optimization and parallelization of the applications for multi-core processors, and to what extent algorithms used by the application are threaded and multi-core ready. Figure 5.3 shows block diagram of basic components of a multi-core processor.

Multi-core processor has two or more cores on a single chip, whereas many-core processor, also called massively multi-core processor, has a lot more cores (tens or hundreds) on a single chip, resulting in advanced performance on highly parallel applications. This concept is realized by Intel MIC (Many Integrated Core) Architecture.

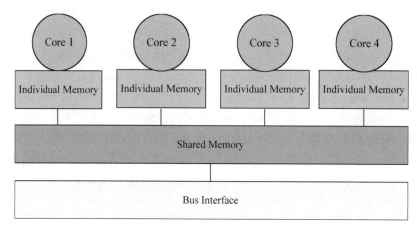

Figure 5.3　A multi-core processor

2. CPU

CPU is an integrated circuit that is considered to be the computing core【核心】 and control center of a computer. Its main function is to execute instructions of the computer and perform operations on data. In most architectures, it is composed of three parts that work together: an arithmetic logic unit (ALU)【算术逻辑单元】, a control unit【控制单元】, and a set of registers【寄存器】(Figure 5.4).

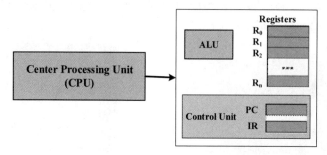

Figure 5.4　Central processing unit (CPU)

Arithmetic Logic Unit (ALU): ALU performs basic arithmetic operations, which include addition, subtraction, multiplication, and division. It also performs logical operations such as AND 【与】, OR 【或】, and NOT【非】. Apart from the two operations, the ALU can also perform comparison and shift operations. We discussed some shift operations on data in Chapter 4, including logical shift operations【逻辑移位操作】, and arithmetic shift operations【算术移位操作】.

Control Unit (CU): The control unit is responsible for selecting, interpreting, and directing execution of the program instructions, and controlling different operations of the computer. It sends and receives control signals to other units like a human brain that controls all the body operations. It handles multiple tasks, such as fetching【获取指令】 and decoding instructions, and managing execution.

Registers: Registers have limited storage capacity and high speed to read or write【寄存器存储能力有限，读写速度快】, which works under the direction of the control unit to store instructions, data and addresses temporarily. A CPU contains multiple registers to facilitate its operations. The three most common categories of registers are data registers 【数据寄存器】, instruction registers 【指令寄存器】 and program counter register【程序计数器】. Data registers are used to store operands, operation results, and intermediate results to reduce the access times of memory【数据寄存器用于存储操作数，操作结果和中间结果，以便减少内存的存取时间】. Instruction register (register IR in Figure 5.4) store the instruction currently being executed, including operations codes, address codes, and address information of instructions【指令集存取存储当前正在执行的指令，包括操作码，地址码和指令的地址信息】. Program counter (PC register, as shown in Figure 5.4, also called Instruction Pointer (IP) register, or Instruction Address Register) is used to store the address of the next instruction that is to be executed once the execution of the current instruction is completed【一旦当前指令执行完毕，程序计数器用来存储下一个将要被执行的指令】. In other words, it always points to the address of the next instruction when the current instruction is executed by the microprocessor.

3. GPU

GPU, occasionally called visual processing unit (VPU)【视觉处理单元（器）】, is a concept relative to CPU. GPU is a specialized electronic circuit designed to rapidly manipulate and alter memory to accelerate the creation of images in a frame buffer intended for output to a display. Modern GPUs are very efficient at manipulating computer graphics, and their parallel structure【并行结构】 makes them more effective than general-purpose CPUs for algorithms where processing of large blocks of data is done in parallel.

GPUs are used in embedded systems【嵌入式系统】, mobile phones, personal computers, workstations and game consoles. It is widely used in graphic processing. We can say that, GPU is the "brain" of graphic card.

GPGPU (General Purpose GPU)【通用计算图形处理器】 is the use of a graphics processing unit to crunch data. CUDA is a widely used platform of GPGPU.

CUDA (Compute Unified Device Architecture)【统一计算设备架构】 is a new parallel computing architecture【并行计算架构】 and programming model that is based on NVIDIA GPU. CUDA enabled GPUs are able to solve complex computational problems more effectively. NVIDIA can bring the power of supercomputing to any workstation【工作站】 or server【服务器】, as well as CPU-based server clusters【服务器集群】. CUDA provides an opening environment for GPU computing. When programmed through CUDA, GPU is regarded as a data-parallel computing (DPC)【数据并行计算】 device capable of running many threads in parallel. In addition, CUDA can allocate and manage operations. GPUs have the characteristics of highly intensive computation【高度密集的计算】 and data parallelism. Therefore, CUDA is suitable for large-scale parallel computing field【CUDA 适用于大规模的并行计算领域】.

In the architecture of CUDA, a GPU is comprised of Streaming Multiprocessors (SMs)【流多处理器】. A multithreaded program【多线程程序】 is divided into blocks【块】 of threads that execute independently and concurrently, so that a GPU with more multiprocessors takes less time to execute the program automatically.

5.1.2 Main memory

Main memory is another major component of a computer (Figure 5.5). It consists of a collection of cells【单元】, each with a unique physical address【物理地址】.

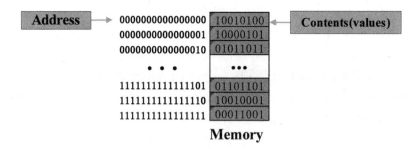

Figure 5.5　Main memory

1. Address space

In memory, each word needs an identifier to be accessed. Although programmers use a name to identify a word (or a series of words), each word is identified by an address at the hardware level. The memory size that any of computer entity occupies is called the address space【任何计算机实体占用的内存大小叫做地址空间】. For example, a memory with 32 kilobytes and a word size of 1 byte has an address space that ranges from 0 to 32768. Table 5.1 shows the units used to refer to memory.

Table 5.1　Memory units

Unit	Kilobyte	Megabyte	Gigabyte	Terabyte
Exact Number of Bytes	2^{10} (1024) bytes	2^{20} (1,048,576) bytes	2^{30} (1,073,741,824) bytes	2^{40} bytes
Approximation	10^3 bytes	10^6 bytes	10^9 bytes	10^{12} bytes

2. Address as bit pattern

Computers operate by storing numbers and address as bit patterns【位模式】, so if a computer has 32 kilobytes (2^{15} bytes) of memory with a word size of 1 byte, then to define an address, you need a bit pattern of 15 bits. In general, if a computer has N words of memory, you need an unsigned integer【无符号整型】 with size of $\log_2 N$ bits to refer to each memory location. Especially, memory address is represented by unsigned binary integers.

Example

A computer has 64 KB (kilobytes) of memory. How many address bits are needed to address any single word in memory?

Solution

The memory address space is 64 KB, or 2^{16} ($2^6 \times 2^{10}$) bytes. This means you need $\log_2 2^{16}$ or 16 bits, to address each word.

3. Memory types

There are two types of memories: RAM (Random Access Memory)【随机存取存储器】 and ROM (Read-Only Memory)【只读存储器】.

In a computer, most common type of memory is RAM. A typical characteristic of RAM is that it is volatile【易失性】: Once the system is powered off, all information in RAM is lost【一旦系统断电，RAM 中的所有信息会丢失】. RAM technology is divided into two classifications: SRAM (Static RAM)【静态随机存储器】 and DRAM (Dynamic RAM)【动态随机存储器】.

The contents of ROM are written by the manufacturer【ROM 存储的内容是由制造商写入的】.

The user can read but not write on ROM【用户只能读取不能写入】. In contrast to RAM, it is nonvolatile: The data is not lost when the power is turned off. The characteristic of ROM is that data cannot be modified or deleted after it is stored. It is used in computer and other electronic devices that do not need frequent updates or changes【它被用于那些不需要频繁更新和改变的计算机和其他电子设备】. There are three types of ROM: PROM (Programmable Read-Only Memory)【可编程只读存储器】, EPROM (Erasable Programmable Read-Only Memory)【可擦可编程只读存储器】and EEPROM (Electrically Erasable Programmable Read-Only Memory)【电可擦可编程只读存储器】.

4. Memory hierarchy

The performance of a memory usually depends on three main indicators: the speed, capacity and price【内存的性能通常取决于三个主要因素：速度、容量和价格】. Computer users need memory that has high capacity, high speed, and low cost. However, it is difficult to meet all the requirements. The solution is using memory hierarchy architecture, mainly including registers【寄存器】, cache memory【高速缓存】, and main memory (Figure 5.6).

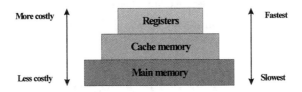

Figure 5.6 Memory hierarchy

The hierarchy is based on the following.

- The registers inside the CPU are very small amount of high-speed memory where speed is crucial.
- Cache is a moderate amount of medium-speed semiconductor memory【高速缓存是一个数量适中的中速的半导体内存】.
- Main memory is a large amount of low-speed memory, and used for accessing data and program that are most frequently used by CPU【主存储器是数量最多的低速内存，用于存取CPU最频繁使用的数据和程序】.

5. Cache memory

The speed of cache memory is faster than the main memory but slower than the CPU/GPU and its registers【高速缓存的速度高于主存储器，但又低于CPU/GPU和它的寄存器】. Cache memory is located between the CPU/GPU and main memory, and is used to bridge the gap between the high clock speed of microprocessor and low access time of memory (Figure 5.7).

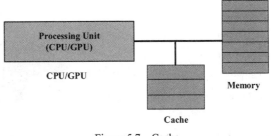

Figure 5.7 Cache

5.1.3 Input/output subsystem

The third component in a computer is the input/output (I/O) subsystem that allows a computer to communicate with the outside world. It consists of two parts: peripherals (I/O devices) and I/O control system. There are two broad categories of I/O devices: non-storage【非存储设备】, and storage devices【存储设备】. Typical characteristics of I/O system are device independence【设备独立性】, asynchronous【异步】I/O, and capability of sending or receiving information in real-time【实时性】.

Non-storage devices allow the CPU/memory to communicate with the outside world, but they cannot store information. This type of devices includes keyboard, monitor, printer, camera, scanner, light pen, handwriting input panel, voice-input device, and plotter【非存储设备包括键盘、监视器、打印机、照相机、扫描仪、光笔、手写输入面板、语音输入设备以及绘图仪】.

Storage devices, although classified as I/O devices, can store huge quantities of information and are sometimes called auxiliary storage devices. Storage devices have lower prices than main memory, and their contents are nonvolatile. Storage devices are classified into two categories: portable storage devices【便携式存储设备】, and built-in storage devices【内置存储设备】(Figure5.8).

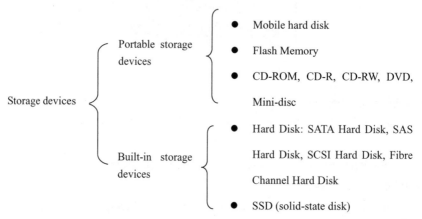

Figure 5.8　The category of Storage devices

1. **Mobile hard disk**

Mobile hard disk【移动硬盘】 uses hard disk as storage medium and can exchange large amounts of data between computers (Figure 5.9). Since the interface of mobile hard disk uses USB and IEEE1394, so it has high-speed transmission【由于移动硬盘采用 USB 和 IEEE1394 接口，所以传输速率很高】. The characteristics of mobile hard disk mainly have 4 aspects.

- Large capacity【容量大】.
- Small volume【体积小】: There are three types of sizes: 1.8inches, 2.5inches and 3.5inches.
- High speed: The transmission rate of USB 2.0 interface is 60 MB/s. However, USB 3.0 interface is up to 625 MB/s.
- Light and convenient【轻便】.

Figure 5.9　Mobile hard disk

2. Flash Memory

Flash Memory is an independent storage medium that is applied in mobile phones, digital cameras, laptop, MP3, and other digital products【闪存卡是一种独立的存储媒介，应用于手机、数码相机、笔记本电脑、MP3 和其他数码产品】. Flash Card uses Flash Memory technology to store electronic information. Memory card is a small, portable and easy to use. In addition, since the memory card has good compatibility, it's convenient to exchange data between different digital products. Flash Memories consists of MMC (Multi-Media Card)【多媒体卡】, SD Card (Secure Digital Card)【安全数码卡】, MS Card (Memory Stick Card)【记忆棒】, CF Card (Compact Flash Card)【小型快速闪存卡】, XD Picture Card【XD 图形卡】, USB flash disk, and Micro Drive (MD)【微型硬盘】. Table 5.2 shows a series of flash memories.

Table 5.2　Flash Memory Cards

Series	Figure	Size	Characteristic	Products	Application
MMC		32mm×24mm×1.4mm	with 7-pin interface; Have no read and write protection switch	RS-MMC MMC PLUS MMC mobile MMC micro	Digital camera, phone and some PDA products (Siemens MP3, phone 6688)
SD Card		32mm×24mm×2.1mm	encryption function and security of data	miniSD microSD T-Flash SDHC	Panasonic and dopod phone, smart phone
MS Card		not unified	With 10-pin interface; Have write protection switch	MS PRO MS Duo MS PRO Duo MS Micro	Media player, digital camera and PDA
CF Card		43mm×36mm×3.3mm	Use flash technology; Low cost; Good compatibility	CF CF+/CF2.0 CF 3.0	digital cameras of Canon and Nikon
XD Picture Card【XD 图形卡】		20mm×25mm×1.7mm	High read and write speed; low power consumption	XD M-XD H-XD	Digital cameras of Fuji

Micro Drive (MD) refers to a very small volume of hard disk data storage device that is like a coin. Compared with the same period of flash memory products, the characteristics of micro disk are considerable memory capacity and very fast read/write speed. However, its typical shortcomings are easy to heat, shorter lifetime and high electricity consumption when operating【微型磁盘典型的缺点是操作时容易发热，使用寿命较短，高电耗】. Micro disk can be built in a number of large-capacity storage requirements of electronic devices such as personal digital assistants (PDA), portable music players, laptops, or even more powerful mobile phones (Figure 5.10).

USB flash disk

USB flash disk (Figure 5.11) is a high-capacity miniature mobile storage product using USB interface. It needs no physical drive and connects with a computer via USB interface to realize plug-and-play function【它不需要物理驱动，是通过USB接口与计算机连接，从而实现即插即用功能】. USB flash disk has a great number of advantages as follows.

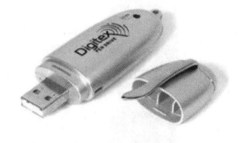

Figure 5.10　Micro drive　　　　　　　　Figure 5.11　USB flash disk

- Portability【便携性】: The USB flash disk is small and easy to carry.
- Big storage capacity: General storage capacity of USB flash disk is 2GB, 4GB, 8GB, 16GB, 32GB, or 64GB.
- Low price.
- Reliable performance.
- Shock and drop resistance【防震抗摔性】: USB flash disk has no mechanical component, so it has strong ability to resist shock.

In addition, the USB flash disk also has resistance to moisture, dust, water, and high and low temperature.

It is known that USB 2.0 and USB 3.0 are a new interface technology in the field of computer, but they have some technology differences as follows.

- The maximum transmission bandwidth【最大传输带宽】: USB 2.0 is 480Mbps (60MB/s), however, USB 3.0 is up to 5.0Gbps (625MB/s).
- Data transmission: USB 3.0 uses full-duplex data transmission and can read and write data simultaneously, but USB 2.0 only supports half duplex【数据传输：USB 3.0使用全双工方式进行数据传输，且能够同时读写，而USB 2.0仅支持半双工】.
- Power supply capability【供电能力】: USB 3.0 is 1A, but USB 2.0 is 0.5A.

3. Mini-disc

Mini-disc is an audio storage media and the diameter is 6.4 centimeters. Figure 5.12 is a mini disc that can record 80 minutes, but only a quarter of the area of a CD【图 5.12 是一张迷你光盘，可以记录 80 分钟，但只有一张 CD 的四分之一大】. For portability, the disc is encapsulated in a plastic housing.

4. SSD (solid-state disk)

A SSD (solid-state disk)【固态硬盘】is a solid-state data storage device that uses integrated circuit assemblies as memory to store data persistently. Figure 5.13 shows a 2.5-inch SSD that can be used in laptops and desktop computers. SSD is composed of control units and storage unit (including Flash chip 【闪存芯片】, DRAM chip). SSDs have no mechanical components【SSD 没有机械构件】. This distinguishes them from traditional electromechanical magnetic disks such as hard disk drives (HDDs) 【硬盘驱动器】 and floppy disks【软盘】, which contain spinning disks and movable read/write heads 【SSD 区别于传统的电子机械式磁盘，如硬盘驱动器，软盘，传统磁盘包含旋转磁盘和可移动的读写磁头】. Compared with electromechanical disks, SSDs are typically more resistant to physical shock【物理震荡】, run silently, have lower access time, and less latency【低延迟】. The advantages of SSD have 6 aspects:

Figure 5.12　Mini-disc　　　　Figure 5.13　A 2.5-inch SSD

- Fast read/write speed: SSD uses flash memory as storage medium, and has a faster read/write speed relative to mechanical hard disk. Besides, SSDs do not have heads, and the seek time is almost zero【SSD没有磁头，寻道时间几乎为零】.
- Shock and drop resistance: SSD does not contain mechanical component, so even in the case of high-speed movement, it still functions normally. What's more, when the collision and shock happen, SSD can reduce the probability of data loss to the minimum.
- Low power consumption【低功耗】: The power consumption of SSD is less than a traditional hard disk.
- No noise: SSD has no mechanical motor and fan, so its noise is 0 dB when working.
- Wide working temperature range: Most SSDs can operate at -10 to 70 degrees Celsius.
- Portability: SSDs are lighter in weight than conventional hard disk.

The disadvantages of SSD mainly have 3 aspects: limited capacity, lifetime limitation, and high price.

5. Hard disk

From an overall perspective, the interface of hard disk 【硬盘】 is divided into four types: IDE (Integrated Drive Electronics)【电子集成驱动器】, SCSI (Small Computer System Interface)

【小型计算机系统接口】, SAS (Serial Attached SCSI)【串行连接 SCSI】 and Fibre Channel【光纤通道】. Therefore, the hard disk consists of SATA (Serial Advanced Technology Attachment)【串行高级技术附件】 Hard Disk, SCSI Hard Disk, SAS Hard Disk, and Fibre Channel Hard Disk (Table 5.3).

Table 5.3 Hard disks

Name	Interface	Characteristic	Application
SATA Hard Disk	SATA	Stronger error correction capability; High data transmission reliability	PC
SCSI Hard Disk	SCSI	Wide range of application; Wide bandwidth; Low CPU utility【CPU 占用率】; Hot-plug【热插拔】	Minicomputer【小型机】
SAS Hard Disk	SAS	Fast speed【转速】; Fast seed【寻道】	Enterprise-class Storage, Servers, Finance industry
Fibre Channel Hard Disk	Fibre Channel	Hot-plug; Remote connection; Wide bandwidth; Large amount of equipment connections	Workstations【工作站】, servers, mass storage sub-network【海量存储子网络】

In addition, nano-disk (Figure 5.14) is also known as quanta disk 【量子磁盘】 and its storage density has reached 41011 bits/inch^2. Magnetic nanowire array film is used as a high-density perpendicular magnetic recording medium【磁性纳米阵列膜可以作为一种高密度的垂直磁记录介质来使用（被称为量子磁盘）】. Perpendicular magnetic recording technology can guarantee great signal-to-noise ratio【垂直磁记录技术可以确保好的信噪比率】.

Figure 5.14 Nano-Disk

5.2 Components Interconnection

In the previous sections, we explored the characteristics of CPU/GPU, main memory, and I/O respectively. The three subsystems are interconnected in a stand-alone computer, and information needs to be exchanged between them. So, this section will analyze how these three subsystems are interconnected.

5.2.1 CPU/GPU and memory connection

CPU/GPU and memory are connected by three groups of bus【总线】: data bus【数据总线】, address bus【地址总线】, and control bus【控制总线】 (Figure 5.15).

（1）Data bus

The data bus is made up of several wires, each carrying 1 bit at a time. The size of the word can decide the number of wires. If the word is 64 bits (8 bytes) in a computer, you need a data bus with 64 wires. Data bus is used to transfer data between memory, CPU/GPU, and peripherals.

（2）Address bus

The address bus permits access to a particular word in memory. The size of address space of memory (that is the CPU's addressing ability) is determined by the number of wires of address bus【内存地址空间的大小由地址总线数决定】. If the address bus carries n bits, the address space of memory is 2^n words.

（3）Control bus

The control bus handles communication between the CPU/GPU and memory. Control bus is mainly used by CPUs for transmitting control signals and timing signals including read/write signals, chip-select signals, interrupt-response signals【中断响应信号】, and so on【控制总线主要用来通过CPU传输控制信号和时序信号，包括读写信号、片选信号、终端响应信号】. If a computer has 2^r control actions, you need r wires for the control bus.

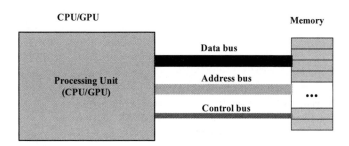

Figure 5.15 Connecting CPU/GPU and memory using three buses

5.2.2 I/O devices connection

Because of the difference between the nature of I/O devices and the nature of CPU/GPU and memory, I/O devices cannot be connected directly to the data buses, address buses, and control buses. Therefore, I/O devices are attached to the buses through I/O controllers or interfaces which is an intermediary to handle this difference. There are some specific controllers that are used to connect input/output devices and buses (Figure 5.16).

A controller has two ways of data communication: serial【串行】, and parallel【并行】. A serial controller has only one wire connection to the device, such as FireWire【火线】, USB 2.0, and USB 3.0. However, a parallel controller has several connections to the device so that multiple bits can be transferred at the same time, such as SCSI.

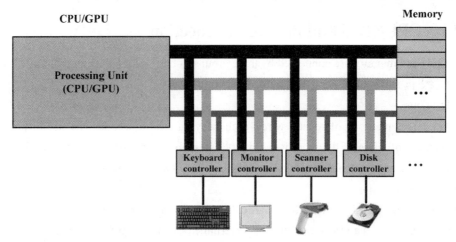

Figure 5.16　Connecting I/O devices to the buses

5.3　Machine Cycle

General-purpose computers process data by executing a set of instructions called a program. Specifically, the CPU repeatedly runs the machine cycle【机器周期】to execute a program. The machine cycle can be simplified into four phases: fetch【取（指令）】, decode【解码】, execute【执行】, and store, as shown in Figure 5.17.

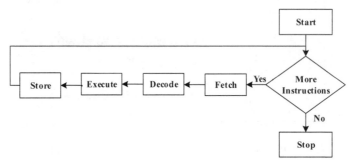

Figure 5.17　The steps of machine cycle

（1）Fetch: The address of the next instruction to be executed is stored in the program counter register【下一个将要执行的指令被存储在程序计数器中】. At the beginning of a machine cycle, the control unit orders the system to copy the instruction into the instruction register in the CPU【在机器周期的开始，控制单元要求系统把指令复制到指令寄存器中】. And then the program counter is increased by one to point to the next instruction in memory【程序计数器加一，并指向内存的下一个指令】.

（2）Decode/Interpret: In the decode step, the control unit decodes the instruction in the instruction register into binary code【二进制码】. The decoded result is used for subsequent operations that the system will perform【在解码阶段，控制单元把指令寄存器的指令解码成二进制码。解码后的结果用于系统接下来将要执行的操作中】.

In order to complete a task, additional memory accesses may be required for the instruction to be executed.

(3) Execute: After the first two steps, the control unit passes the decoded information as a sequence of control signals to relevant components of the CPU to perform the corresponding operations.

(4) Store: The results generated by the operations are stored in the main memory, or sent to an output device【操作生成的结果被储存在主存储器或发送到输出设备】. PC register could be updated to point a new address from which the next instruction will be fetched【程序计数器将被更新，并且指向下一个指令将被获取的新的地址】.

5.4　Computer Architectures

Computer architecture【计算机架构】 is the conceptual design and fundamental operational structure of a computer system. It can be defined as the science and art of selecting and interconnecting hardware components to create computers that meet functional, performance, and cost goals. In this section, we discuss two common computer architectures.

CISC【复杂指令集计算机】 (Complex Instruction Set Computer) represents a computer using a full set of computer instructions to operate【CISC 代表一种使用整套计算机指令执行的计算机（架构）】. Several low-level operations can be performed by single instructions. Programming CISC-based computers is easier because single instructions are capable of multi-step operations. Examples of CISC instruction set architectures【指令集架构】 are PDP-11, VAX (Virtual Address eXtender to PDP-11), Motorola 68k, and x86.

RISC【精简指令集计算机】(Reduced Instruction Set Computer) represents a CPU design strategy to have a small set of instructions which perform a minimum number of simple operations. Well-known RISC families include DEC Alpha, AMD 29k, ARC, ARM, Atmel AVR, Intel i860 and i960, MIPS, Motorola 88000, PA-RISC, Power (including PowerPC), SuperH, and SW26010.

Although both RISC and CISC try to make balance in the architecture, operation and running, software and hardware, compile-time and run-time to become more efficient; they use different methods, they have many differences, as follows:

(1) Instruction system【指令系统】: RISC designers focus on those instructions that are frequently used and try to make simple and efficient features【RISC 设计者关注那些经常使用的指令，尽量使它们具有简单高效的特色】. When implementing the special functions on RISC machines, it may be less efficient【当在 RISC 机器上实现特殊功能时，效率可能较低】. However, the instruction system of CISC has specific instructions to perform specific functions, so it has higher efficiency to handle special tasks【CISC 指令系统有专用指令来完成特定的功能，因此，处理特殊任务时效率更高】.

(2) Memory operations【存储器操作】: RISC has restrictions on memory operations and makes the control operations simple. But CISC has multiple memory manipulation instruction【CISC 有多个存储器指令】.

(3) Program: The assembly language programs of RISC usually require large memory space【RISC 汇编语言程序通常需要大的存储空间】. The program is complex and time-consuming when realizing

some special functions. The assembly language programs of CISC are relatively simple and complex programs are easier, so CISC is more effective.

（4）Interrupt: RISC can respond to interrupt【响应中断】 in the appropriate place where an instruction is being executed, however, CISC responds to interrupt after an instruction is executed【RISC 机器在一条指令执行的适当地方可以响应中断，而 CISC 机器是在一条指令执行结束后响应中断】

5.5　References and Recommended Readings

For more details about the subjects discussed in this chapter, the following books are recommended:

- Cragon H G. Computer Architecture and Implementation. Cambridge: Cambridge University Press, 2000.
- Englander I. The Architecture of Computer Hardware, Systems Software, and networking: An Information Technology Approach. Hoboken, NJ: Wiley, 2009.
- Ercegovac M D, Lang T, Moreno J H. Introduction to Digital Systems. Hoboken, NJ: Wiley, 1998.
- Hamachar C, Vranesic Z, Zaky S. Computer Organization. Boston: McGraw-Hill, 2002.
- Hennessy J L, Patterson D A. Computer Architecture: A Quantitative Approach. Waltham, MA: Elsevier, 2012.
- Mano M M. Computer System Architecture. Upper Saddle River, NJ: Practice Hall, 1993.
- Null L, Lobur J. Computer Organization and Architecture. Sudbury, MA: Jones and Bartlett, 2003.
- Stallings W. Computer Organization and Architecture. Upper Saddle River, NJ: Prentice Hall, 2002.
- Warford J S. Computer Systems. Sudbury, MA: Jones and Bartlett, 2010.
- Fu H H, Liao J F, Yang J Z, et. al. The Sunway TaihuLight supercomputer: system and applications. Science China Information Sciences, 59(7): 072001: 1-16, 2016.
- Gasior G. SSD prices in steady, substantial decline: A look at the cost of the current generation. The Tech Report, June 21, 2012.
- Kasavajhala V. SSD vs. HDD Price and Performance Study. Dell PowerVault Technical Marketing, May 2011.
- NVIDIA Corporation. CUDA C Programming Guide. http://www.nvidia.com, February 2014.
- Prinslow G: Overview of Performance Measurement and Analytical Modeling Techniques for Multi-core Processors. http://www.cse.wustl.edu/~jain/cse567-11/ftp/multcore/index.html, 2011.
- STEC. SSD Power Savings Render Significant Reduction to TCO. Retrieved October 25, 2010.
- "What is a Solid State Disk?" Ramsan.com, Texas Memory Systems, archived from the original on 4 February 2008.
- Whittaker Z: Solid-state disk prices falling, still more costly than hard disks. Between the Lines. ZDNet. Retrieved 14 December 2012.

5.6　Summary

- A computer consists of CPU, main memory, and input/output components.
- The processor consists of uniprocessor, and multi-core processor. Uniprocessor has only one processor per system. However, multi-core processor refers to an integrated circuit to which two or more complete cores are attached into a processor.
- Multi-core processors have enormous advantages of computing performance and can meet the requirements of multi-tasking processing and computing at the same time【多核处理器具有巨大的计算性能优势，并且能满足同时进行多任务处理和计算的需求】。
- With floating point speed of 93 petaflops per second (or 93 quadrillion floating point operations per second), Sunway TaihuLight is twice as fast and three times as energy-efficient as China's Tianhe-2.
- Each node in Sunway TaihuLight has one SW26010 processor (or chip) that produces speeds of 3.06 teraflops/s with 260 cores and uses ShenWei-64 instruction set.
- CPU is an integrated circuit that is considered to be the computing core and control center of a computer. It has three parts: an arithmetic logic unit, a control unit, and a set of registers. ALU performs arithmetic, logic, and shift operations on data. Registers are fast stand-alone temporary storage locations on the processor. The control unit controls the operations of each part of CPU.
- GPU is a specialized electronic circuit designed to rapidly manipulate and alter memory to accelerate the building of images in a frame buffer intended for output to a display【GPU是一个专用电子电路，目的在于快速操纵和改变存储以便加速帧缓冲区中图像的构建，从而输出到显示器上】。
- GPGPU (General Purpose GPU) is the use of a graphics processing unit. CUDA is a widely used platform of GPGPU.
- CUDA (Compute Unified Device Architecture) is a new parallel computing architecture that is based on NVIDIA GPU【CUDA是一个基于NVIDIA GPU的新的并行计算架构】。
- Main memory is a collection of storage locations. Two types of memory are available: RAM, and ROM.
- In a computer, most common type of memory is RAM. RAM technology is divided into two categories: SRAM and DRAM.
- The contents of ROM are programmed by the manufacturer. The user can read but not write on ROM. There are three types of ROM: PROM, EPROM, and EEPROM.
- Computers need high-speed memory for registers, medium-speed memory for cache memory, and low-speed memory for main memory.
- Since the interface of mobile hard disk uses USB or IEEE1394, it has high speed transmission. For the maximum transmission bandwidth of USB 2.0 is 480Mbps (60MB/s), however, USB 3.0 is up to 5.0Gbps (625MB/s).
- The maximum transmission bandwidth of USB 20 is 480Mbps (60MB/s), however, USB 3.0 is

up to 5.0Gbps (625MB/s).
- Flash Memory is an independent storage medium that is applied in mobile phones, digital cameras, laptops, tablets, MP3 players, and other digital products.
- Flash Memories consists of MMC (Multi-Media Card), SD Card (Secure Digital Card), MS Card (Memory Stick Card), CF Card (Compact Flash Card), XD Picture Card, USB flash disk, and Micro Drive.
- Compared with electromechanical disks, SSDs are typically more resistant to physical shock, have lower access time, faster transfer speeds, and less latency.
- Hard disk drive (HDD) includes SATA hard disk, SCSI hard disk, SAS hard disk, and Fibre Channel hard disk, and nano-disk.
- CPU/GPU and memory are connected by three groups of bus: data bus, address bus, and control bus.
- The CPU repeatedly runs the machine cycle to execute a program. A simplified cycle can consist of four phases: fetch, decode, execute, and store.
- The two designs for CPU architecture are CISC and RISC.

5.7 Practice Set

1. What are the three components that form a computer?
2. Define the following terms: _____.

a. Uniprocessor

b. Multi-core processor

c. Many-core processor

3. What are the components of a CPU?
4. What is the function of the ALU?
5. Describe the function of the main memory.
6. What is a GPU? And what is the difference between CPU and GPU?
7. Registers are used to temporarily store instructions, data, and addresses. List three most common registers.
8. Compare and contrast RAM and ROM.
9. What are the advantages and disadvantages of SSD?
10. What are the characteristics of USB flash disk?
11. What is the purpose of cache memory?
12. List four types of flash disks and compare them with each other.
13. What is the difference between USB 2.0 and USB 3.0?
14. Introduce four types of hard disks and analyze their characteristics.
15. Compare and contrast the CISC architecture and the RISC architecture.
16. The _____ is a computer component that performs operations on data.

  a. CPU   b. memory   c. I/O hardware   d. none of them

17. _____ is a unit that can add two inputs.
    a. An ALU   b. A register   c. A control unit   d. A tape drive
18. _____ is located between the CPU/GPU and main memory.
    a. Cache   b. Controller   c. Register   d. None of them
19. If the memory address space is 32 MB and the word size is 4 bits, then _____ bits are needed to access each word.
    a. 12   b. 24   c. 26   d. 28
20. Which of the following are input/output devices? _____.
    a. CPU   b. keyboard   c. printer   d. monitor
21. There are _____ bytes in 8 terabytes.
    a. 2^{13} b. 2^{20} c. 2^{43} d. 2^{30}
22. The _____ controller is a parallel interface and daisy-chained connection for I/O devices.
    a. FireWire   b. SCSI   c. USB 3.0   d. IDE
23. If the memory has 2^{24} words, the address bus needs to have _____ wires.
    a. 8   b. 24   c. 32   d. 2^{24}
24. A control bus with eight wires can define _____ operations.
    a. 8   b. 16   c. 256   d. 512
25. The CPU runs the machine cycle one by one until the program ends, and the machine cycle can be simplified into three phases in the specific order: _____.
    a. Fetch, execute, and decode   b. Decode, execute, and fetch
    c. Fetch, decode, and execute   d. Decode, fetch, and execute
26. A computer has 32 MB of memory. Each word is 8 bytes. How many bits are needed to address each single word in memory?

Unit 6
Computer Networks

In this chapter we mainly describe networks. Computers are often connected to form a network, and the networks are connected to form an internetwork【互联网络】. We divide the chapter into the following parts: types of networks, TCP/IP model, and devices in networks. At last, we introduce the new development in networks.

6.1　Types of Networks

We classify networks by their spanning distance【跨距】 as shown in Figure 6.1. At the top are the personal area networks that supports one person instead of a group. Beyond the personal area networks are longer-range networks. These can be classified as local, metropolitan, and wide area networks, each with increasing scale. Finally, the connection of multiple individual networks is called an internetwork. Without doubt, the worldwide Internet is the best-known (but not the only) example of an internetwork 【全球互联网是互联网络的最著名的（但不是唯一的）例子】.

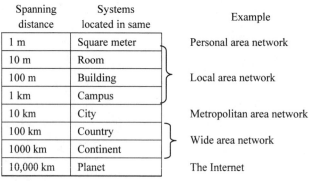

Figure 6.1　Classification of networks by spanning distance

6.1.1　Personal area network

A personal area network (PAN)【个人局域网】 is a technology that let devices to communicate with other nearby devices【与其他临近的设备进行交互】 near a single individual person, over short ranges typically within a range of 10 meters. Personal area networks can be constructed with cables or be wireless. For example, a computer can be connected with its peripherals【周边设备】, or a laptop can be connected with a smart phone or a printer, or a headset can be connected to a mobile phone【耳麦可以连接到手机】, using some form of wireless technology. PANs can be wired or wireless. Wired PAN may be linked together using USB and FireWire. Wireless PAN (WPAN) is based on the standard IEEE 802.15【无线个人局域网是基于 IEEE 802.15 的标准】and typically use ANT+, Bluetooth【蓝牙技术】, some infrared connections, ZigBee【紫蜂协议】, UWB (Ultra Wideband)【超宽频】, RFID (Radio Frequency Identification)【无线射频识别】, 6LoWPAN, or NFC (Near Field Communication)【近距离无线通信技术】.

ANT+ is proprietary wireless sensor network technology used in the collection and transfer of sensor data. It is used in sports watches, workout machines, cycling power meters, fitness equipment, thermometers, speedometers, calorimeters, blood pressure monitors, blood glucose meters, heart rate monitors, position tracking, temperature sensors, and so on. Bluetooth, also known as IEEE 802.15.1, uses short-range radio waves【短程无线电技术】and are energy-efficient for small devices. Infrared (IR) connections【红外线连接】use infrared light to represent 0s and 1s. IR is widely used for consumer remote controls【用户遥控】. ZigBee, also known as IEEE 802.15.4, is a low-cost, low-power technology, which makes it particularly well-suited for remote monitoring and controlling applications【远程监控和控制应用程序】such as wireless lighting control and fans, home and building automation and wireless security systems, traffic management systems, medical and industrial sensor applications, radio remote control toys, games, and so on. UWB enables movement of massive files at high data rates over short ranges, such as wireless monitors, wireless printers, camcorders, and PC peripherals. RFID is based on the concept of sending a query over a radio wave and receiving its subsequent reply【通过无线电波发送一个查询和接受其后续答复】. RFID is used in smart cards, tracking systems, library books, and supply chain management. 6LoWPAN means IPv6 over Low-Power Wireless PANs. It is an open standard defined by the Internet Engineering Task Force (IETF). It is a new technology that enables the integration of IPv6 and low power devices that conform to IEEE 802.15.4 standard in a PAN, and is primarily used for home and building automation, and healthcare and environmental monitoring. NFC is an ISO/IEC set of standards and involves an initiator and a target. Many smart phones use NFC to make payment, and to integrate credit card.

6.1.2 Local area network

A local area network (LAN)【局域网】is designed to share resources between computers【在电脑之间进行资源共享】. It can be defined as a combination of computers and peripheral devices (e.g. printers【打印机】) connected through a transmission medium (e.g. cable【电缆】). Figure 6.2 shows three common topologies【拓扑结构】belonging to LANs.

In the bus topology【总线拓扑结构】, terminal devices like computers are connected through a common medium called a bus. In this topology, all computers receive the frame【数据帧】and check its destination address【目的地址】when a station sends a frame to another computer. The frame is accepted and the data contained in the frame is processed if the destination address in the frame header matches the physical address of the station【数据帧头中目的地址与站点的物理地

址相匹配】; otherwise, it is discarded. The second topology is the star topology 【星形拓扑结构】 in which computers are connected via a switch (an intricate hub that controls the forwarding of the frame) or a hub【集线器】 (a device that facilitates connection). The LAN works logically like a bus in an Ethernet environment using a hub. The hub just sends data out of all its interfaces【把数据从所有的端口发送】. But using a switch, the address in the frame is checked by the switch, and the frame is sent out only through the interface of the destination【数据帧仅仅通过目的端口进行发送】. Finally, we introduce the ring topology【环形拓扑结构】. In this topology, when a computer needs to send a frame to another computer, it sends it to its logical neighbor in the ring【把数据发送到环线结构的逻辑邻居中】. Here the frame is regenerated and sent to the next neighbor until it reaches the final destination. The destination opens the frame, copies the data, either removes the frame from the ring or adds an acknowledgement【添加一个确认】 and sends the frame (via the ring) back to the original sender. In the latter case, the frame is then removed by the sender.

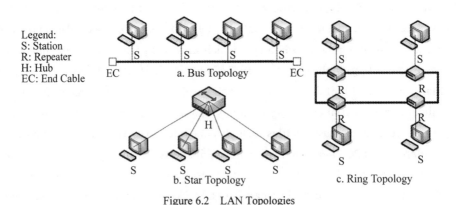

Figure 6.2　LAN Topologies

6.1.3　Metropolitan area network

A metropolitan area network (MAN)【城域网】 is the computer communication network build across a city. It belongs to broadband LAN【属于宽带局域网】. MAN's transmission medium is optical fiber【光纤】. The main function of MAN is regarded as internet backbone which connects different computers, databases and LANs【连接不同电脑，数据库，局域网的主干网】. The MAN has a wide range of applications including interactive personality TV【交互式网络电视】, remote medicine【远程医疗】, remote education【远程教育】, remote monitoring【远程监控】, and so on.

6.1.4　Wide area network

A wide area network (WAN)【广域网】 is the connection of different countries or areas which is also called remote computer network (RCN)【远程网】 to achieve resource sharing. By far, the biggest WAN covers 180 different countries including China. And it grows at a rate of 15% per month. WAN also includes wireless wide area network【无线广域网】 (WWAN), a large WAN that uses cellular network technologies. WWAN are commonly referred to as 3G/4G/5G networks【无线广域网通常被认为是包括3G/4G/5G 网络】.

6.1.5 Internetwork

As mentioned earlier, you can connect individual LANs, MANs and WANs (using routers or gateways) to form a network of networks, called an internetwork or an internet. Today, there are many private and public internets. However, the most famous is the Internet. The Internet was originally a research internetwork designed to connect several different heterogeneous networks【连接几个不同的异构网络】. It was sponsored by the defense advanced research project agency (DARPA). Today, the Internet is an internetwork that connects millions of computers throughout the world.

6.2 TCP/IP Model

The OSI (Open Systems Interconnection)【开放式系统互联】 reference model【参考模型】 and the TCP/IP (Transmission Control Protocol/Internet Protocol) reference model【TCP/IP 参考模型】 are the two important network architectures. Although the protocols associated with the OSI model are not used any more, the model itself is actually quite general and still valid. The TCP/IP model has the opposite properties: the model itself is not of much use but the protocols are widely used【模型本身并没有大量使用，但是它的协议被广泛使用】. For this reason, we will mainly look at the TCP/IP reference model in detail.

6.2.1 Layers of TCP/IP Model

The TCP/IP【传输控制协议/互联网协议】 is a stack of protocols【协议栈】 that officially controls computers to connect the Internet. But due to the widespread use of TCP/IP protocols, OSI model is just a theoretical model. The structure of OSI model is as follows: (Figure 6.3).

1. Physical layer and data-link layer

Physical layer【物理层】 and data-link layer【数据链路层】 are responsible for specifying the physical medium, the signal, and the bits. In other words, they just provides services to the upper layer protocols【它们仅仅为上层协议提供服务】.

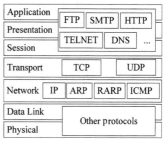

Figure 6.3 TCP/IP and OSI model

2. Network layer

Network layer【网络层】 (or more accurately, the internetwork layer or the internet layer) is located in the third layer in TCP/IP model which is one of the most important protocols. The main functions of network consists of route selection【路由选择】, congestion control【拥塞控制】, and error detection【误差控制】. The network layer has several protocols including Internet Protocol (IP)【互联网协议】, Address Resolution Protocol (ARP)【地址解析协议】, Reverse Address Resolution Protocol (RARP)【逆地址解析协议】, and Internet Control Message Protocol (ICMP)【控制报文协议】. Using these protocols, network layer can pack the data from the upper layer【包装来自上层协议的数据】 and provide reliable point-to-point (PPP) service【提供可靠的点对点服务】.

3. Transport layer

At the transport layer【传输层】, TCP/IP defines two important protocols: Transmission Control Protocol (TCP)【传输控制协议】, and User Datagram Protocol (UDP)【用户报文协议】. Its main functions are segmentation and reassembly of packets【数据包分段和重组】, addressing according to port number【根据端口号寻址】, connection management【连接管理】, and flow control【流控制】. TCP is distinguished from UDP as follows: TCP is just connection-oriented protocol【面向连接的协议】; it can provide reliable service by establishing "three-way handshake" between the client and the server【在客户端和服务器之间建立三次握手】; after establishing "three-way handshake", it can ensure the accuracy of data and the order of the data【数据的准确性和数据的顺序】. On the contrary, UDP is connectionless-oriented protocol【无连接协议】; UDP programming is always simple without "three-way handshake" so it may not provide reliable service and cannot assure the accuracy of the data. Both TCP and UDP are providing end-to-end service【端到端的服务】.

4. Application layer

The TCP/IP application【应用层】layer is defined to transfer data between different applications in two computers【在两台计算机的不同的应用程序之间进行数据传输】. It is located at the top of TCP/IP model. There are many protocols in application layer, such as FTP, SMTP, Telnet, HTTP, and so on. The definition of these protocols is as follows:

- File Transfer Protocol (FTP)【文件传输协议】 is the protocol which can be used to transfer different files. In general, the server sends data to the client which is called "download"【服务器向客户端发送数据称为"下载"】 and the client sends data to the server which is called "upload"【客户端向服务器发送数据称为"上传"】. The transmission mode of FTP has two ways: ASCII transmission mode and binary data transmission mode【ASCII传输模式和二进制数据传输模式】. FTP protocol is one of the most important protocols in the application layer.

- Simple Mail Transfer Protocol (SMTP)【简单邮件传输协议】 is a protocol which can ensure to send and receive email【发送和接收邮件】. It is a relative simple protocol. When a user specifies a message to a receiver or several receivers, the message can be transmitted. We can usually use telnet program to test the SMTP server.

- Telnet (Terminal Network)【远程登录协议】 is mainly used for Internet session. According to this protocol, it may start a session with username and password【通过用户名和密码开启一个会话】 to login into the server. So the user can send instruction to server【传输指令到服务器】, as in the console controlling the server.

- Hypertext Transfer Protocol (HTTP)【超文本传输协议】 is an object-oriented protocol 【面向对象的协议】. It is used to send data from WWW server to local browser. It can help the computer get the data including text, multimedia files, audio, video, images, and so forth quickly.

6.2.2 IP address (IPv4 & IPv6)

An Internet Protocol address (IP address)【互联网通信协议地址】is a numerical label assigned to each device connecting to a computer network that communicates through the use of Internet Protocol. It must be noted that IP address【网际协议地址】is unique in the Internet and each device always owes one IP address.

1. IPv4

IP address in IPv4 consists of 4 bytes (32 bits). The maximum number of Internet addresses in IPv4 is 2^{32}. TCP/IP model uses a notation that is more convenient to read called dotted-decimal notation【点分 10 进制计数法】 taking place of the binary form. Figure 6.4 shows an example which transfers the binary to dotted-decimal notion.

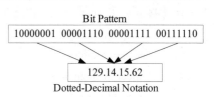

Figure 6.4　IPv4 address in dotted-decimal notation

2. IPv6

The Internet addresses in IPv4 tend to deplete because of limited Internet addresses【因为 IP 地址是有限的，分配不均等原因，渐渐的 IP 地址被耗尽】, uneven distribution, etc., and cannot satisfy the rapid growth of Internet users. IPv6 was proposed by Internet Engineering Task Force (IETF)【互联网工作任务组】 due to the rapid exhaustion of IPv4 address space. The main features of IPv6 are discussed below.

First and foremost, IPv6 has four times longer addresses than IPv4. They are 128 bits long, which solves the problem that IPv4 can not solve: providing a huge effectively number of Internet addresses【大量有效的互联网地址】. The second major improvement of IPv6 is the simplification of the header. It contains only seven fields (versus 13 in IPv4). This change allows routers to process packets faster【这种改变能够使得路由器可以更快地处理数据包】and thus reduce processing time by intermediate nodes and improves throughput and delay. The third major improvement is better support for options. This change is essential with the new header because previously required fields are now optional (because they are not used so often). In addition, each of the extension headers in IPv6 is optional and represented in different way, making it simple for routers to bypass options which are not intended【绕过一些不需要的选择】. This feature speeds up packet processing time. A fourth area in IPv6 is in security. IETF had a newspaper story about precocious 12-year-olds using their personal computers to break into banks and military bases【入侵银行系统和军事基地】 all over the Internet. Finally, IPv6 can achieve quality of service (QoS) as it offers a packet prioritization feature that provides improved response time of real-time applications.

6.2.3　Wireless LANs

In the mid-1990s, the industry considered that a wireless LAN【无线局域网】 standard might be very helpful idea, so the IEEE committee【电气与电子工程委员会】 accepted a task which drawn up a wireless LAN standard. Then it made the first decision. Other LAN standards include 802.1, 802.2, and 802.3, up to 802.10, therefore the wireless LAN standard was given the name of 802.11. It has a common name—WiFi. Because the standard is very important, we still call it 802.11.

The remaining task was harder. The initial task was to seek an appropriate frequency band【寻找一个切当频带】 that was not already in use, preferably across worldwide. The method taken was the opposite of that used in mobile phone networks. In place of expensive spectrum licenses, 802.11 systems work in unlicensed bands, for example, the ISM (Industrial, Scientific, and Medical) bands defined by ITU-R (e.g., 902-928 MHz, 2.4-2.5 GHz, 5.725-5.825 GHz). All devices are permitted to utilize this spectrum in case they limit their transmit power to allow different devices coexist【通过限制传输速率

达到不同设备的共存】. Without doubt, this implies that 802.11 radios might be competing with cordless phones, garage door openers, and microwave ovens. 802.11 networks are composed of infrastructure and wireless clients. Infrastructure usually consists of wireless routers called APs (Access Points)【接入点】 that are installed in buildings, and wireless clients can be laptops, smartphones, or workstations equipped with wireless NICs (Network Interface Cards)【网卡】. Access points also referred to as base stations【基站】. The access points plugs into the wired network, and enables all clients to connect through an access point. There is another mode of operation for IEEE 802.11 which is called an ad hoc network【自组织网络】. This mode allows direct communication between the clients that are in radio range, for example, communication between two computers in a building without an access point. However, compared with access point mode, it is used much less. Both modes are shown in Figure 6.5.

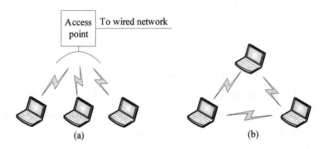

Figure 6.5　(a) Wireless network with an access point　(b) Ad hoc network

Mobility becomes another issue. If a mobile client is moved away from the access point【远离接入点】 it is using to another; there must be some way of handing it off. The solution is that an 802.11 network may include multiple cells, each with its own access point, and a distribution system that links all the cells. We often use switched Ethernet【交互式以太网】 to realize the distribution system, but other technology may be used for it. When the mobile clients move, they can use other access point with a better signal to replace the original access point and shift their association【代替原来的接入点和转变它们的关系】. As a whole, we can consider the entire system as a single wired LAN.

Finally, there comes security issues. Since wireless networks broadcast【广播】 transmissions, so nearby devices who want to get the information can easily receive the packets not intended for them. In order to stop this situation, a well-known encryption scheme named as WEP (Wired Equivalent Privacy)【有线等效保密】 was assigned to the 802.11 standard. The aim of this was to make the wireless security like that of wired security【使得无线能够像有线一样安全】. With the development of technology, WiFi Protected Access【无线保护访问】, initially called WPA, but now upgraded to WPA2, has replaced the old schemes.

The 802.11 standard is a revolution in wireless networking. Beyond buildings, it can be used in trains, planes, boats, and automobiles. People are able to surf the Internet whenever they want or wherever they go. ZigBee technology is one of the most important wireless network standard【紫蜂技术是最重要的无线网络标准之一】. The characteristics of ZigBee are short transmission distance, low power consumption, low speed, low complexity, and low cost【短传输距离，低功耗，低速度，低复杂度，低成本】. It can help mobile phones and all manner of consumer electronics, from game consoles to digital cameras, to communicate more easily.

6.3 Devices in Networks

Nowadays there are a lot of connecting devices to connect different networks. However, connecting devices can be divided into four types based on their functionality: repeaters, bridges, routers, and gateways. Repeaters and bridges are devices that connects different LANs【中继器和网桥用来连接不同的局域网】. Routers and gateways are devices that connects different WAN and MAN【路由器和网关用来连接不同的广域网和城域网】.

6.3.1 Repeaters

A repeater【中继器】 is a device that is used to forward physical signals between two different network nodes【在两个不同的网络节点中传递物理信号】. Repeater mainly works at the physical layer to make a strong, clean copy of the signal, by adjusting the original signal and regenerating it.

The main features of repeater are as follows: expanding the connecting distance【扩展连接距离】; increasing the maximum number of network nodes【增加最大的网络节点数】; using different transmission rates between different network segments【在不同的网段中使用不同的传输速率】; improving the reliability and performance of LAN【改进局域网的性能和可靠性】.

6.3.2 Bridges

Bridge【网桥】 is an equipment which works at the data link layer to achieve LAN interconnection【实现局域网互联】. It can divide a big LAN into several network segments. In general, bridge receives MAC data frame from LAN. After opening, and validating the frame, bridge will send it to the physical layer according to another LAN reassembling data frame【根据另外一个局域网重组数据帧】. Because the bridge is a data link layer device, it may not deal with the data header that upper layer protocols add.

The basic features of bridge are as follows: bridge can connect different LANs at data link layer; bridge can connect LANs which have different data link layer protocol【不同的数据链路层协议】, different transmission medium and different transmission speed【不同的传输媒介和不同的传输速率】; bridge connect the internet and transmit/receive data, gives information about the storage, filter MAC address, and retransmit data packets【数据包重发】.

6.3.3 Switches

Switches【交换机】 are widely used in our daily life like laboratory in college, office in company, or ordinary house. A switch is a device that can amplify the internet through providing more connecting port for subnet. The main features of switch are high flexibility, high performance-to-price rate, relative simplicity【高灵活性、高性价比、相对简单】 and easy implementation. Since Ethernet networking technologies have become the most popular in LANs, so network switch has become the most popular switch. On the whole, switch consists of WAN switch and LAN switch.

6.3.4　Routers

Router【路由器】 is the most important device in the internet because it can be used to connect bigger internet like MAN or WAN. Routers mainly use two types of routing protocols namely static routing protocols, and dynamic routing protocols【静态路由协议和动态路由协议】. The main functions of router are path selection according to routing protocol【根据路由协议进行路径选择】, packet switching【分组交换】, building the map of the network in the form of routing table【以路由表的形式建立网络地图】, routing data between networks【网络间路由数据】, accessing control lists【访问控制列表】, restrict broadcasts to the LAN【局域网限制广播】, and so on. It mainly works at network layer (OSI Model's layer 3).

Router has five types including access router, enterprise router, backbone router, terabit router, and multi-homing router. And we can choose appropriate router according to different demands. Figure 6.6 shows an example of a router that connects different networks.

6.3.5　Gateways

Gateway【网关】 is also called protocol converters that connects network above the network layer. It is the most complex of all internet connecting device. A firewall and a proxy server acts as a gateway【防火墙和代理服务器都能作为网关】. Some people regard the gateway as a router that consists of protocol translators, and others regard a router as a gateway because a router can control the path through which packets of data are sent in and out. The distinction between the two terms is disappearing.

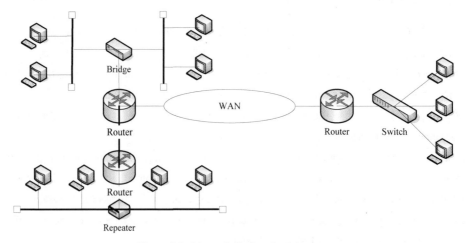

Figure 6.6　Network devices in an internet

6.4　New Development in Networks

6.4.1　Third, fourth, and fifth generations of mobile networks

Over the past 40 years, the architecture of the mobile phone network【移动网络的体系结构】 has realized tremendous growth, and this growth still continues exponentially. Mobile network is also called cellular network【蜂窝网络】. In the first-generation mobile phone systems【在第一代移动网络系统中】, analog traffic channels were used to transmit voice calls instead of digital bits. In 1982, AMPS (Advanced Mobile Phone System)【高级移动电话系统】 was widely used in the United States and it belongs to the first generation system. In order to increase capacity, improve security, and provide text messaging, in the second-generation mobile phone systems digital technology was used to transmit voice calls. Since 1991, GSM (Global System for Mobile communications)【全球移动通信系统】 is the most widely deployed mobile phone system designed as a second generation (2G) wireless telephone technology.

The third generation (3G) systems were initially implemented in 2001 and provide both broadband digital data services and digital voice【宽带数字数据服务和数字语音】. They also come with a number of different technologies and many different standards to choose from. 3G is an ITU specification providing bandwidth of at least 2 Mbps for stationary or walking users and 384 kbps in a moving vehicle. UMTS (Universal Mobile Telecommunications System)【通用移动通信系统】 is a 3G wireless standard that is widely deployed worldwide and uses WCDMA (Wideband Code Division Multiple Access)【宽带码分多址移动通信系统】 as the radio transmission standard. It offers maximum 14 Mbps in the downlink and almost 6 Mbps on the uplink. 4G and 5G technology is the extended technique of 3G which can increase the data transmission speed and add some new communication technology. 5G technology is the proposed next major phase of mobile telecommunications standards【移动通信标准】 beyond the current 4G/IMT-Advanced standards. 5G planning includes some surprising capabilities such as ultra-low latency, broader coverage, higher efficiency, more frequency spectrum, high bandwidth, Internet connection speeds faster than current 4G, networked robots connected to the cloud in real time, delivering uninterrupted communication flow to self-driving cars, Internet of Things (IoT), smart cities, remote surgeries, and other improvements. The Next Generation Mobile Networks (NGMN) Alliance feels that 5G will be rolled out by 2020 to meet business and consumer demands. China has gone through the periods of "2G tracking, 3G breakthrough, 4G synchronization"【2G 跟踪，3G 突破，4G 同步】. To promote 5G, China takes the lead in setting up the IMT-2020 (5G) Promotion Group in the Asia-Pacific Region. China also integrates the elite force from industry, learning, research, and application【整合产、学、研、用精锐力量】 to actively communicate with the International Telecommunication Union (ITU). In addition, some telecommunication operators and industry manufacturers【运营商、设备制造商】 from China have fully participated in the global mainstream R&D teams including 5GPPP, NGMN Alliance, etc. In summary, China aims to gain superiority of 5G technology【占据 5G 技术制高点】 and lead the development of telecom industry. Future wireless networks will employ multiple antennas and radios to offer faster connection speeds for users.

Table 6.1 shows the comparison of several mobile networks.

The architecture of a cellular network is very different from that of the Internet. It contains several parts, as shown in the simplified UMTS architecture in Figure 6.7. In the first place, it has the air interface which can be used to communicate between the cellular base station and the mobile device【蜂窝基站和移动设备】 (e.g., the cell phone). In the past decades, the technological advancements in the air interface have enormously increased wireless data rates. The UMTS air interface is based on Code Division Multiple Access (CDMA)【码分多址】.

Table 6.1 Comparison of several mobile networks

Item/networks	2G	3G		4G	5G
	GSM	TD-SCDMA	CDMA2000	TD-LTE-Advanced	
Speed	5kbps	384kbps	2.4Mbps	1Gbps	1Gbps
Bandwidth	200kHz	1.6MHz	2.5MHz	100MHz	28GHz
Security	Low	High	High	High	High
Advantages	Good compatibility; Strong anti-interference; Complete coverage	Avoid breathing effect【避免呼吸效应】; High spectrum efficiency	low construction cost; update from CDMA 1X directly	Effective support for the new band and high-bandwidth applications; A substantial increase in peak data rate	Easy to manage; Provide consistent connection; High transmission rate
Disadvantages	Poor business performance; Do not support 3G services	Interference problem; High synchronization requirements	Near-far effect【远-近效应】; Multiple access interference【多址干扰】	Hard marketing; Covering regional synergy【覆盖区域协同】	Have not been used; High expenses
Standard Contributor	Europe	China	America	China	None

The radio access network【无线电接入网络】 contains cellular base station and its controller. This section is the wireless side of the mobile phone network. The Radio Network Controller (RNC)【无线网络控制器】 manages how the spectrum is utilized. The base station implements the air interface. It is a temporary tag and we can call it Node B.

The remaining part of the mobile phone network is the core network【主干网络】 which transmits the traffic for the radio access network. The UMTS core network evolved from the core network used for the 2G GSM system that came before it. However, in the UMTS core network, something surprising is occurring.

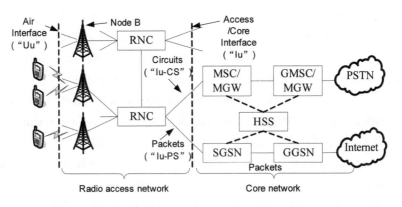

Figure 6.7 Architecture of the UMTS 3G mobile phone network

The surprise in Figure 6.7 is that there are both packets and circuit switched equipments in the core network. This reveals that in mobile phone companies, the mobile phone network in transition can achieve one or sometimes both of the alternatives. In the traditional phone network, older mobile phone networks utilized a circuit-switched core to carry voice calls. It is the legacy of the UMTS network with the MSC (Mobile Switching Center)【移动交换中心】, GMSC (Gateway Mobile Switching Center)【网关移动交换中心】 and MGW (Media Gateway)【媒体网关】 elements which configure links over a circuit-switched core network, for instance, the PSTN (Public Switched Telephone Network)【公用开关电话网络】.

Mobile data-based services in GSM system have evolved considerably, starting with services such as SMS (Short Message Service)【短信服务】, voice telephony, and GPRS (General Packet Radio Service)【通用分组无线业务】. These earlier data services ran at tens of kbps, but users wanted more. Newer cellular phone networks transmit packet data at rates of multiple Mbps. In order to compare, a voice call is carried with the speed of 64 kbps, typically 3–4 times less with compression.

The UMTS core network nodes link directly to a packet-switched network【直接连到分组交换网络】 in order to transmit all the data. The GGSN (Gateway GPRS Support Node)【通用分组无线业务网关支持节点】 and the SGSN (Serving GPRS Support Node)【通用分组无线业务支持点】 transform data packets to and from mobiles and interface to external packet networks【包交换网络】, for example, the Internet.

This transition is also existing in the mobile phone networks that are now being planned and deployed. Internet protocols are even used on mobiles to configure links for voice calls over a packet data network, in the manner of voiceover-IP. IP and packets are used all the way from the radio access to the core network. Of course, the way that IP networks are designed is also changing to support better quality of service. If it did not, then problems with chopped-up audio and jerky video would not impress paying customers.

Mobility is another difference between cellular networks and the traditional Internet【移动性是蜂窝网络与传统互联网的另一个区别】. When users move between cellular base stations, the packets must be re-routed from the old to the new base station. The technique is called handoff or handover.

Security is another serious issue taken into account by cellular companies due to the need of billing for services and prevent fraudulent payments and claims【防止欺诈付款与索赔】. Unfortunately, that is not enough. Nevertheless, in the evolution from 1G to 4G technologies, mobile phone companies have been able to roll out some basic security mechanisms【运行一些基础的安全机制】 for mobiles.

Cellular networks are destined to play a significant role in future networks. They not just offer the ability to make voice calls, but also offer a number of mobile broadband applications【移动宽带应用】, and this has potential implications for the core network architecture, security of the future networks, and the air interfaces. LTE (Long Term Evolution), a 4G wireless broadband technology, is on the drawing board, even the 3G ongoing design and deployment continues to improve. There are other wireless techniques which provide broadband Internet access to fixed-location clients and mobile clients, such as 802.16 networks under the common name of WiMAX (Worldwide Interoperability for Microwave Access). It is hard to predict entirely what will be the next wave of wireless standards and what will happen to WiMAX and LTE.

6.4.2 GRID

A grid【网格】 computer is a system which can cluster multiple number of same class of computers together【把许多同类型的电脑聚在一起】. A grid computer is linked through a super-fast network and share the devices【设备共享】 such as disk drives, printers, RAM, and mass storage. Grid computing is an efficient scheme with respect to distributed supercomputing OS which has the capability of parallelism. In order to get a common goal, and to solve a single task, grid computing can group computers together from multiple administrative domains and then disappear just as quickly【网格计算能够把多管理领域的电脑聚在一起，用完之后能够快速归还】. One of the major techniques of grid computing is to utilize middleware to divide and apportion pieces of a program【程序划分】 among multiple computers, sometimes up to many thousands. Grid computing contains computation in a distributed fashion, which may also contain the aggregation of large-scale cluster computing-based systems【大规模集群计算系统的聚合】. The size of a grid may vary from small-confined to a network of computer workstations within a company, for example, to large, public collaborations across several corporations and networks. The concept of confined grid is also called an intra-nodes cooperation【节点内协作】, whereas the concept of a larger, wider grid is related to an inter-nodes cooperation【节点间的协作】. Grids are a form of distributed computing whereby a "super virtual computer"【超级虚拟计算机】is made up of many networked loosely coupled computers which can act together to accomplish very large, complicated tasks. The technique has been widely used in computationally intensive scientific【计算密集型科学】, mathematical【数学】, and academic problems【学术问题】 through volunteer computing, and it is used in commercial enterprises for various applications such as economic forecasting【经济预测】, seismic analysis【地震分析】, drug discovery【药物研发】, and back office data processing in support for e-commerce and Web services【后台数据处理对电子商务和网络服务的支持】. Resource coordination【资源协调】 on grids become increasingly complex, especially when coordinating the flow of information across distributed computing resources. Grid workflow systems【网格工作流系统】 are a specialized form of a workflow management system. These systems are particularly designed to compose and execute a series of computational or data manipulation steps, or a workflow【执行一系列的计算或者数据操作步骤或者是一个工作流】, in the Grid context. In short, distributed or grid computing is a special style of parallel computing【并行计算】 which depends on entire computers (with onboard CPUs, storage, power supplies, network interfaces, etc.) linked to a network (public, private or the Internet) by a conventional network interface, such as Ethernet. The idea is very different from the conventional notion of a supercomputer【这与传统的超级计算机的思想是非常不一样的】, which contains several processors linked by a local high-speed computer bus.

6.4.3 RFID and Sensor Networks

Networks made up of computing devices, such as mobile phones and laptops, are easier to identify. RFID uses radio-frequency electromagnetic fields【移动电话微波电磁场】 to automatically identify people or everyday objects and can connect them to more powerful computers through networks.

An RFID tag is typically of the size of a postage stamp【一个RFID标签的大小与一张邮票的大小一样】. It can be embedded into or affixed to any object so that it can be tracked by RFID readers【无线辨识读卡器】 anywhere. The object can be anything such as a cow, a passport, a car tire, an aircraft part, or a book. The tag is made up of a tiny microchip【微芯片】 containing a unique identifier【唯一标识符】 for objects that have it and an internal antenna that receives radio transmissions. RFID readers (also called interrogators) acts like access points for RFID tagged objects which can discover tagged items when they come into a defined interrogation point or range and interrogate them for their information as shown in Figure 6.8. Applications include checking identities, managing the supply chain【供应链管理】, timing races【计时比赛】, identifying theft, tracking files, tracking assets, tracking people, checking inventory, access control, protecting electronics, managing livestock【畜禽管理】, managing contactless payment systems, and replacing barcodes【更换条形码】.

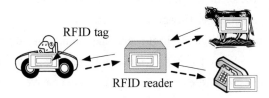

Figure 6.8 RFID used to network everyday objects

RFID frequency ranges are LF (Low Frequencies), HF (High Frequencies), UHF (Ultra-High Frequencies), and SHF (Super-High Frequencies).

UHF RFID【超高频无线射频识别】 is one of the most common forms. It is used in wide range of applications, including drivers' licenses, shipping pallets【运输托台】, and so on. UHF RFID operate at 860-960 MHz frequency bands depending on the country of use. UHF RFID frequency range in China is 840-928 MHz. RFID Reader always transmits encoded radio signals to interrogate the tag. Tags communicate at distances of several meters by changing between reflecting reader signals and absorbing them; the reader is able to decode tag signals【解码标签信号】 from tag by filtering out the signal it is transmitting. This kind of operating is known as backscatter【反向散射】.

HF RFID is another popular kind of RFID. It operates at frequency range of 13.56 MHz and is likely to be in your passport, contactless payment systems【非接触支付系统】, books, and credit cards. HF RFID has a short range, typically up to a meter, because the physical mechanism is based on near-field coupling and magnetic induction rather than backscatter.

LF RFID is other form of RFID using frequency around 132 kHz. It is now replaced by HF RFID and used for animal tracking. It is likely to be in your cat.

RFID readers have to deal with the multiple access problem【多重接入问题】 within reading range. This means that when a tag hears a reader or the signals from multiple tags it may not simply respond. If multiple tags simultaneously transmit then tag collisions【标签冲突】 can occur. Multiple access collisions are: tag-tag collision, reader-reader collision, and reader-tag collision. The solution is inspired by the approach taken in 802.11: before replying with their identification, tags wait for a short random interval, which permits the reader to decrease individual tags and interrogate them further.

Another problem is privacy and security. The ability of RFID readers to easily track an object, and hence the people who utilizes it, is an invasion of individual privacy【个人隐私侵入】. Unfortunately, securing RFID tags is challenging because they lack the storage, power, circuitry, computational, and communication resources necessary to run strong cryptographic algorithms【加密算法】. Tags cannot store long passwords or calculations in their working memory. Instead, they are able to store a short password of perhaps only 32-128 bits. If an identity card can be remotely read by an official at a border, what is to prevent the same card from being tracked by other people without your knowledge? Not much.

A step up in capability from RFID is the sensor network【传感器网络】. Sensor networks offer the ability to monitor aspects of the physical world. Up till now, they have been widely used for scientific experimentation, for instance, zebra migrations【斑马迁徙】, monitoring bird habitats, monitoring environmental information, and volcanic activity【火山活动】. They are also used for military surveillance. However, the business applications, such as tracking of frozen, refrigerated, or otherwise perishable goods, healthcare, and monitoring device for vibration does not seem so far behind.

Sensor【传感器】nodes are small-sized computers which operate in infrastructure-less environment, consume little energy, relatively inexpensive, and can monitor physical or environmental conditions, such as vibration, temperature, pressure, sound, and so on. Many nodes are placed in the environment that is to be monitored. Typically, they have batteries or an embedded form of energy harvesting such as scavenging energy from vibrations or the sun. Other components of a sensor node are a microcontroller【单片机】, transceiver with an antenna, external memory【包含天线, 外部存储器的收发器】, electronic circuit, and one or more sensors. As with RFID, a key challenge is power consumption constraints, and the sensor node must carefully deliver its sensory data to the collection point. A common approach is to self-organize the nodes to relay messages【通过自组织节点传递消息】 for each other, as shown in Figure 6.9. It is known as a multi-hop network【多跳网络】.

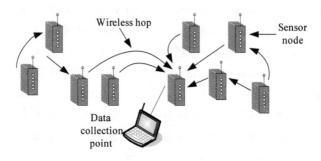

Figure 6.9   Multi-hop topology of a sensor network

6.4.4   Software Defined Network

Software defined network (SDN)【软件定义网络】 refers to an approach for computer network virtualization that allows network administrators【网络管理员】 to programmatically configure applications and network services through abstraction of lower-level functionality. This is done by decoupling【解耦】 the system that makes decisions about where traffic is sent (the control plane【控制层】) from the underlying systems that forward traffic to the selected destination (the data plane【数据层】).

Figure 6.10 shows a high-level overview of the software-defined networks architecture.

The following list defines and explains the architectural components.

- SDN Applications are the procedures which can communicate their network requirements to the SDN Controller through the northbound interface (NBI)【北向接口】.
- The SDN Controller is a logically centralized controller or network operating system【软件定义网络控制器是一个逻辑中央控制器或者是一个网络操作系统】. The main functions of SDN Controller are translating the requirements from the SDN Application layer and relays them to the SDN Datapaths【SDN数据通路】 and introducing the SDN Applications with new abstractions in networking.

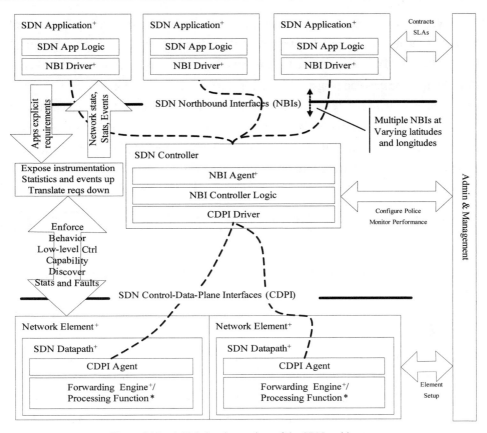

Figure 6.10 A high-level overview of the SDN architecture

- The SDN Datapath is a logical network device. It can expose uncontended control over its advertised forwarding and data processing capabilities. The logical representation may encompass all or a subset of the physical substrate resources.
- The SDN CDPI (Control-Data-Plane-Interface)【控制数据平台接口】is an interface between an SDN Controller and an SDN Datapath. It provides programmatic control of all forwarding operations, capabilities advertisement, statistics reporting, and event notification.
- SDN NBIs (Northbound Interfaces) are interfaces between SDN Applications and SDN Controllers. They provide abstract network views and enable direct expression of network behavior and requirements.

6.4.5 Microwave signal transmission technology and the gigabit optical fiber

Microwave signal transmission technology via optical fiber【微波信号光纤传输技术】combines the advantages of microwave signal and optics technology. It contains four parts: electrical to optical converter, optical to electrical converter, microwave driver, and optical cable【电/光转换器件、光/电转换器件、微波驱动器件以及光缆组成】. The former two parts is used to accomplish the optical to electrical exchange function of microwave. The function of microwave driver is to drive the microwave to a proper level or modulation. And the optical cable is the transmission medium of modulated optical signal. It has two modulation modes: direct modulation mode, and external modulation mode【直接调制模式和外调制模式】. The former mode has the advantage of relatively simple technical solution but suffers from limited transmission ranger. The latter mode is contrast to the former mode.

The gigabit optical fiber【千兆光纤】 and electronic media converter is Ethernet electrical to optical converter between the electrical signal of Gigabit Ethernet and the optical signal, which accord with the IEEE802.3z/ab standard. It obtains a good stability and reliability. Its major function is the interconnection of Ethernet backbone network and metropolitan area network.

6.5 References and Recommended Readings

- Comer D E. The Internet Book: Everything You Need to Know about Computer Networking and How the Internet Works. 4th Edition, Prentice Hall, 2006.
- Crovella M, Krishnamurthy B. Internet Measurement: Infrastructure, Traffic and Applications. NY: John Wiley & Sons, 2006.
- Forouzan B, Mosharraf F. Foundations of Computer Science. Cengage Learning EMEA, 2008.
- Nadeau T D, Gray K. Software Defined Networks. CA: O'Reilly Media, 2013.
- Tanenbaum A S, Wetherall D J. Computer Networks. Boston: Prentice Hall, 2011.
- Fan G. Design of the gigabit optical and electronical media converter based on BCM5421S. Manufacturing Automation, 01, 2013.
- Jing-guo W, Shao-lin Z. Microwave signal transmission technology via optical fiber, Optical Communication Technology, 2009.
- http://www.baike.com/wiki/TD-CDMA.
- https://en.wikipedia.org/wiki/Software-defined_networking.
- SDN Architecture Overview, https://www.opennetworking.org/images/stories/downloads/sdn-resources/technical-reports/SDN-architecture-overview-1.0.pdf.
- Ted Rappaport: NYU Wireless' Rappaport envisions a 5G, millimeter-wave future – FierceWireless Tech, www.fiercewireless.com. 2016.
- Wikipedia: Grid Computing, http://en.wikipedia.org/wiki/Grid_computing, Nov. 2012.

6.6 Summary

- Networks can be divided into PANs, LANs, MANs, WANs, and internetworks.
- PANs can be wired or wireless. Wired PAN may be linked together using USB and FireWire. Wireless PAN is based on the standard IEEE 802.15 and typically use ANT+, Bluetooth, some infrared connections, ZigBee, UWB, RFID, 6LoWPAN, or NFC.
- Three common topologies belonging to LANs are bus, star, and ring topology.
- MAN belongs to broadband LAN. Its transmission medium is optical fiber.

- WWAN is a large WAN that uses cellular network technologies. WWAN are commonly referred to as 3G/4G/5G networks.
- The OSI reference model and the TCP/IP reference model are the two important network architectures.
- TCP/IP model contains physical and data-link layer, network layer, transport layer, and application layer.
- An Internet Protocol address (IP address) is a numerical label which is assigned to each device connected to the Internet. IP address in IPv4 consists of 4 bytes (32 bits). IPv6 has four times longer addresses than IPv4. They are 128 bits long.
- A repeater is a device that is used to forward physical signals between two different network nodes.
- Bridge is data communication device that works at data link layer to achieve LAN interconnection. It transmits and receives MAC data frame from LAN and validates the frame.
- Switch is a device that amplifies the signal along the internet by providing more connecting ports for the subnet.
- Functions of routers are selecting the best route according to routing protocol, connecting multiple networks, and accessing control lists.
- Gateways connect multiple digital devices using different connection-oriented transport protocols, understand the format and contents of the data, and translates the contents of a message from one format to another.
- Third/fourth/fifth generations of mobile networks are the new development in networks. The architecture of the mobile phone network has realized tremendous growth. It also brings new opportunities and challenges. 2G, 3G, 4G, 5G are the generations of wireless technologies.
- A grid computer consists of multiple number of computers which have same class and are clustered together. It is connected through a super-fast network and share the devices like disk drives, mass storage, printers, and RAM.
- Radio Frequency Identification (RFID) uses radio-frequency electromagnetic fields to automatically identify people or everyday objects and can connect them to more powerful computers through networks.
- Wireless sensor networks offer the ability to monitor aspects of the physical world. Sensor nodes are small-sized computers which operate in infrastructure-less environment, consume little energy, relatively inexpensive, and can monitor physical or environmental conditions, such as vibration, temperature, pressure, sound, and so on.
- Software defined network (SDN) refers to an approach for computer network virtualization that allows network administrators to programmatically configure applications and network services through abstraction of lower-level functionality. The SDN consists of the following components: SDN Application, SDN Controller, SDN Datapath, SDN Control to Data-Plane Interface, and SDN Northbound Interfaces.
- Microwave signal transmission technology via optical fiber combines the advantages of microwave signal and optics technology. The gigabit optical and electronic media converter is

Ethernet electrical to optical converter between the electrical signal of Gigabit Ethernet and the optical signal, which accord with the IEEE802.3z/ab standard.

6.7 Practice Set

1. Give the definition of common physical network topologies.
2. Define four types of networks.
3. Distinguish between point-to-point and multipoint connection.
4. List the technologies for wireless PANs.
5. Name the layers of the OSI Model.
6. What is the purpose of FTP, SMTP, Telnet, and HTTP?
7. Write the main function of the transport layer in the TCP/IP protocol suite and list the addresses used in this layer.
8. Write the main function of the network layer in the TCP/IP protocol suite and list the addresses used in this layer.
9. Differentiate between a frame and a packet.
10. List type of connecting devices and their functions.
11. List the technical aspects of 2G/3G/4G/5G wireless technologies.
12. What is the purpose of RFID? How does it work?
13. Compare and contrast the low-, high-, ultra-high, and super-high frequencies.
14. What is software defined network? What components does it consist of?
15. List the four parts of microwave signal transmission technology via optical fiber.
16. The OSI model consists of _____ layers.
    a. five   b. six   c. seven   d. none of these
17. The _____ layer of the OSI model organizes bits from the physical layer into logical data units called frames.
    a. physical   b. data-link   c. network   d. transport
18. The _____ layer of the OSI model controls how a person on the network gains access to the data.
    a. data-link   b. transport   c. application   d. physical
19. The _____ layer of the OSI model compresses data.
    a. physical   b. data-link   c. session   d. presentation
20. _____ is a protocol used for transferring files.
    a. FTP   b. SMTP   c. TELNET   d. HTTP
21. _____ is a protocol that can provide mail services.
    a. FTP   b. SMTP   c. TELNET   d. HTTP
22. _____ is a protocol that can access and transfer documents on the WWW.
    a. FTP   b. SMTP   c. TELNET   d. HTTP
23. A _____ document has fixed contents.

a. static   b. dynamic   c. active   d. all of these

24. The transport layer protocol of TCP/IP is called _____.
    a. TCP   b. UDP   c. IP   d. a and b

25. _____ is also known as IEEE 802.15.1
    a. Bluetooth   b. RFID   c. NFC   d. TCP/IP

26. Change the following IP addresses from dotted-decimal notation to binary notation:_____.
    a. 114.32.7.28   b. 129.4.16.8   c. 208.3.54.12   d. 38.33.2.1
    e. 255.255.255.255   f. 230.8.6.0   g. 230.8.31.255   h. 255.240.0.0

27. Change the following IP addresses from binary notation to dotted-decimal notation:_____.
    a. 01111110   11110001   01100111   11111111
    b. 01010111   11011100   11100000   00000101
    c. 00011111   10000100   00111111   11011101
    d. 10001111   11110101   00110111   00011101
    e. 11110111   10010011   11100111   00001111
    f. 11111111   11110000   00000000   00000000

Unit 7
Operating Systems

A modern computer is a complex system, which consists of many components, such as at least one processing element, main memory【内存】, disks【硬盘】, network interfaces【网络接口】, and various input/output devices【输入/输出设备】. Managing all these components is an exceedingly challenging job. For this reason, computers are equipped with a layer of software called the operating system (OS)【操作系统】, whose job is to provide user programs【用户程序】 with a better, simpler, cleaner model of the computer and manage all the resources just mentioned.

Roles of an Operating System

An operating system is system software that manages computer hardware【硬件】 and software【软件】 resources and provides common services for computer programs (Figure 7.1). It presents application programs with elegant and consistent abstractions【抽象】 to work with and provides more functionality【功能性】 than the hardware itself does (Figure 7.2).

Operating systems have two mainly unrelated functions:
- Provide a clean abstract set of resources.
- Manage hardware resources.

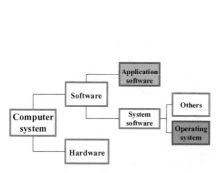

Figure 7.1 Computer system

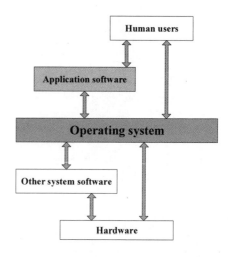

Figure 7.2 Roles of an operating system

7.1 Categories

7.1.1 Batch systems

Batch system【批处理系统】 is a kind of operating system that executes a series of programs automatically. It has been associated with mainframe computers【大型计算机】 in early 1950s. The processing【处理过程】 of a typical computer in the 1960s and 1970s was managed by a human operator【操作员】. In earliest systems, the user's program was stored as a deck of【一副】 punched cards【穿孔卡片】 and delivered to the operator to be executed. The computer operator would return the card deck and printed results to the user, perhaps the next day. Each program to be executed was called a job【作业】. During this era, operating systems were very simple; they only ensure that all of the resources were transferred from one job to the next.

7.1.2 Time-Sharing systems

Before the introduction of time-sharing systems【分时操作系统】, it is important to introduce a new term: process【进程】. A job is a program to be run【运行】, a process is an instance【实例】 of a program that is being executed and waiting for resources.

In order to use system resources efficiently, multiprogramming【多道程序】 was introduced. It allows more than one program to run concurrently【并行地】 over a certain period of time. The computer performs multitasking【多任务】 by executing segments of multiple tasks in an interleaved way.

Time-sharing is achieved by the idea of multiprogramming. Time-sharing systems divide the available processor time between multiple processes, which is called time-slice【时间片】. It means each process can be interrupted before it has reached completion. Meanwhile, computer resources could be shared between different processes, with each process being allocated a portion of time to use a resource.

Multiprogramming and time-sharing considerably improved the efficiency of a computer system. But it also required a more complex operating system.

7.1.3 Personal systems

When personal computers (PCs)【个人计算机】 were introduced, there was a need to design an operating system that was user friendly. Since then, operating systems such as DOS (Disk Operating System), Windows (Windows XP, Windows Vista, Windows 7, Windows 8, Windows 10, etc.), MacOS, some distributions【发行版】 of Linux (Ubuntu, Fedora, etc.) were designed and introduced. These operating systems are stored in the computer's hard disk【硬盘】.

7.1.4 Parallel systems

To solve large problems speedily and efficiently, parallel systems【并行操作系统】 were designed to carry out many calculations simultaneously. These systems have multiple CPUs on the same machine. Each CPU can be used to serve one program or a part of a program, which means that many tasks can be accomplished concurrently instead of sequentially【串行地】. These operating systems are more complex than those that have only single CPU. As power consumption【耗电量】by computers has attracted a lot of attention recently, parallel systems (mainly in the form of multi-core processors【多核处理器】) have become the dominant paradigm in computer architecture.

7.1.5 Distributed systems

A distributed system【分布式操作系统】 consists of a collection of autonomous【自治的】 computers, connected through a network and distribution middleware【分布式中间件】. It enables activity coordination of computers and relies on message passing to share the resources of the system. Distributed systems have three significant features: concurrency【并发性】, no global clock【无全局锁】, and independent failures【故障独立性】. In distributed system, one problem may be divided into many sub-problems【子问题】, each of which is solved by one or more computers. Distributed systems combine attributes of the previous generation with new jobs such as controlling security【安全控制】.

7.1.6 Real-time systems

A real-time system【实时操作系统】 is expected to do a task as data comes in, typically without buffering delays【缓冲延迟】. A real-time system is one that must respond to input immediately such as, mission critical applications must be real-time. An example of a real-time embedded system is anti-lock brakes【防锁死刹车】 and security system in a car. Usually real-time response times【响应时间】 are a matter of milliseconds and sometimes microseconds. In contrast, a non-real-time system is one that has no absolute deadline and that cannot guarantee a response time in any situation, even if a fast response is the usual result.

There are two types of real-time systems: hard real-time system【硬实时操作系统】, and soft real-time system【软实时操作系统】.

7.1.7 Network operating systems

Network operating systems (NOS)【网络操作系统】 are operating systems running on the computers that act as servers【服务器】, and are used to connect and coordinate clients that are networked or linked together. Both peer-to-peer and client-server networking models use network operating systems. NOS aims at communication【通信】 and resource sharing【资源共享】 between clients in the same network (such as local area network (LAN)【局域网】 or private network【专用网络】). Typical network operating systems include UNIX, Linux, MAC OS X Server, Netware, Windows Server, and so on.

7.1.8 Embedded operating systems

Embedded operating systems【嵌入式操作系统】 are usually used by embedded devices. Examples of embedded devices are smartphones【智能手机】, tablets【平板电脑】, set-top boxes【机顶盒】 and other small electronics. In some cases, embedded operating system refers to a universal application【泛用型程序】 with specific application software.

7.2 Components

Today's operating systems are very compact, complex, and extremely efficient by design. In order to make the different parts of a computer work together, most operating systems provide certain basic abstractions called components【组件】 to manage different resources in a computer system. It is implemented by abstracting several managers at the top level. Each manager is designed for managing the specific resources, and needs to cooperate with other managers. A modern operating system has at least four types of system components: memory manager【内存管理器】, process manager【进程管理器】, CPU scheduler【CPU 调度器】, device manager【设备管理器】, and file manager【文件管理器】. Of course there are other components such as user interface【用户界面】 (Figure 7.3).

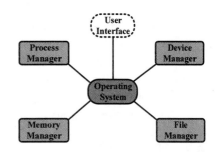

Figure 7.3 Components of an operating system

7.2.1 Memory manager

Main memory holds executing programs and data referenced by those programs. While the memory size of computers has increased tremendously in recent years, the size of the programs and data to be processed has also increased much faster. How to fairly allocate space for those programs and data has become a critical problem. Therefore, as a part of the operating system, memory manager was introduced to efficiently manage the computer memory at system level and prevent the "running out of memory【内存不足】" fault. Its key functions are as follows: keep track of which parts of memory are currently being used and which are free, dynamically allocate【分配】 memory space to processes when they need it, and release 【释放】 it when they are done. In the following, we will discuss several different memory management schemes, from the simple to the sophisticated.

1. Monoprogramming

In the early days, computers had no memory abstraction. In monoprogramming【单道程序】, only one program at a time can be running and most of the memory capacity is dedicated to this program (Figure 7.4). Under this condition, the whole program is loaded into memory before being executed. When the program finishes execution【执行】, the program area is replaced by another program. Obviously, this scheme has several major drawbacks. First of all, it is difficult to run multiple programs at once. Second, the size of program and data must fit in memory.

2. Multiprogramming

Problems of "no memory abstraction" scheme have to be solved to allow multiple programs to be executed at the same time. Therefore, a memory abstraction named "address spaces【地址空间】" was introduced. In multiprogramming【多道程序】, address space is a set of addresses that a program can use to address【编址】 memory. Each program has its own address space which is independent of other programs. Multiprogramming creates a type of abstract memory for programs to live in. Since the 1960s, multiprogramming has gone through several improvements from partitioning【分区调度】to paging【分页调度】or segmentation【分段调度】, and from nonswapping【不可交换】to swapping【交换】(Figure 7.5).

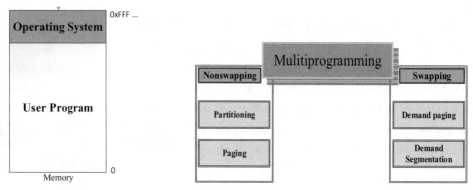

Figure 7.4　Monoprograming

Figure 7.5　Categories of multiprogramming

3. Virtual Memory

As the swapping emerges, virtual memory【虚拟内存】is to be true. Its basic concept is that each process has its own address space, which is broken up into small units called pages【页】(Figure 7.6). The operating system manages the virtual addresses and maps【映射】them into real physical addresses【物理地址】. With one important note, not all pages have to be in the physical memory unless these pages are referenced. Virtual memory combines the memory and disk to form a large range of consecutive address spaces. By using demand paging【请求分页调度】or segmentation or both, virtual memory has been widely used in the modern operating systems.

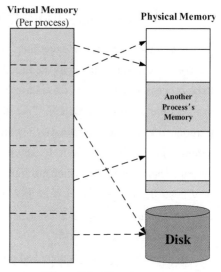

Figure 7.6　Virtual memory

4. Memory Management

Partitioning

The first technique used in multiprogramming is called partitioning. It is the means for dividing the main memory into multiple partitions, usually with contiguous and variable length of address spaces (Figure 7.7). Each partition might contain all the information for a specific program. It begins with allocating a partition to one program and ends with unallocating it. Partitions may be either static【静态的】or dynamic【动态的】, depending on specific operating system. The CPU switches between programs during execution. It may be interrupted by either the instruction of the input/output operation or the expiry of time allocated for this program. Then the CPU saves the address of the memory location where the last instruction was executed and moves to the next program. The CPU repeats the same procedure for the remaining programs. But one of the critical drawbacks of partitioning is external fragmentation【外部碎片】, which occurs in dynamic memory allocation 【动

112

态内存分配】 when empty holes (unused space) is scattered throughout memory.

Paging

Paging improves efficiency of partitioning and prevents external fragmentation. It breaks the memory into fixed-size units called frames【帧】 and the program into pages of the same size. The memory management unit (MMU)【内存管理单元】 is a computer hardware used to map pages to frames. With this technique, there is no need for the program to be contiguous in memory. Two consecutive pages can occupy non-consecutive frames in memory. But the whole program still needs to be in memory before being executed (Figure 7.8).

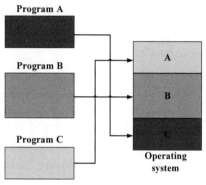

Figure 7.7　Partition

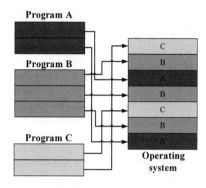

Figure 7.8　Paging

Demand Paging

Demand paging has removed the restriction of storing the entire program in memory. It follows that the pages should occupy the memory only if the executing process needs them. It is a method of virtual memory management and implemented by page table【页表】. The demand paging is like a paging system with swapping where program is in secondary storage and swapped into memory during execution (Figure 7.9).

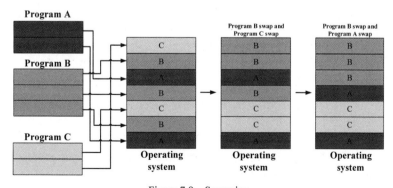

Figure 7.9　Swapping

Segmentation

In paging, a program is divided into fixed-size pages, which is not convenient enough for programmer due to the modularization【模块化】 of programs (a program usually consists of main program and subprograms). Segmentation is a technique that partition memory into logically related【逻

辑相关的】units called segments【段】. It is implemented by segment table【段表】. Different segments usually have different lengths. This scheme allows better access protection【访问保护】 because the hardware will not permit the program to access memory not defined for that specific segment.

Pure segmentation can be implemented without paging. The combination of demand paging and segmentation can further improve the efficiency of the system. A segment may be too large to fit into any available empty space in memory. Memory can be divided into frames, and a module/subprogram can be divided into pages.

7.2.2 Process manager

A principal responsibility of an operating system is process management. Before the discussion of process management, it is necessary to introduce some terminologies.

1. Program, job, and process

A program【程序】 is a set of instructions【指令】 stored on disk or other storage medium【存储介质】. It is non-active【非活跃的】 and may or may not become a job. A job【作业】 is a program from the moment a program is chosen for execution until it has finished running and becomes a program again. The job may or may not run in this duration. A process【进程】 is a program that is being executed. In other words, it is a job that has started execution but has not finished. Figure 7.10 clarifies the relationship between a program, a job and a process.

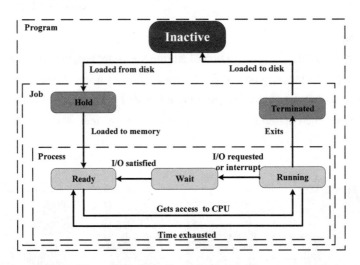

Figure 7.10   State diagram of process manager

The operating system must manage resource allocation to processes, enable information sharing and exchange【信息共享与交换】between processes, protect resources of one process from others, and enable process synchronization【进程同步】.

2. Schedulers

Process manager uses two *schedulers*【调度器】 to change the state of a job or process.

Job scheduler

This part is simple. A job scheduler【作业调度器】 is responsible for process creation from a job and process termination【终止】 (Figure 7.10).

Process scheduler

The process scheduler【进程调度器】 is the component of the operating system that is responsible for deciding which process among the set of runnable processes can run and whether the currently running process should continue running and, if not, which process runs next (Figure 7.10). According to actual situations, there are four events that may occur:

- It switches a process from the running state【运行状态】 into the waiting state【等待状态】 when the process is waiting for some event to happen.
- It switches a process from the running state into the ready state【就绪状态】 if the allotted【分配的】 time has expired.
- It switches the process from the waiting state into the ready state when the event has occurred.
- It switches the process from the ready state into the running state when the CPU is ready to run the process.

3. Job/Process control block and queues

Job/Process control block【作业/进程控制块】: Each job/process is associated with a job/process control block, which stores information about that job/process and is used by process manager to schedule process. Instead of the job or process itself, the process manager stores the job/process control block in the queues【队列】.

Queues: To handle【处理】 multiple processes and jobs, the process manager uses queues. An operating system can have several queues: Job queue, ready queue, and I/O queue.

4. Process synchronization

The main idea for process management to allocate resources for different processes is synchronization【同步】. As long as the resources are used by more than one process, we may encounter two possible problems: deadlock【死锁】, and starvation【饿死】.

Deadlock

Before giving a formal definition of deadlock, we will give an example first. There are two processes. Both of them want to copy a scanned document on a DVD. Process A requests【请求】 permission to use the scanner and get it at once. Process B requests the DVD recorder first and gets it too. Now A requests for the DVD recorder, but B has not release it, so the request cannot be granted. Meanwhile, B asks for the scanner, but the request is denied【拒绝】 until A releases it. In this situation, deadlock occurs because both processes will remain block【阻塞】.

Deadlock does not always occur. It may occur when the operating system does not put resource restrictions【资源限制】 on processes.

Starvation

The relationship between deadlock and starvation is cause and effect 【因果关系】; starvation is caused by deadlock. Starvation occurs when a process is permanently blocked from gaining access to necessary resources due to too many resource restrictions on a process. For example, there are several processes waiting for printing task【打印任务】. Assume the file length is variable, and the resource allocation strategy for printing task of operating system is shortest-file-first 【最小文件优先】. Then, if the processes with short-file printing task never blocks, the processes with large-file printing task will never be scheduled.

7.2.3 CPU Scheduler

In multiprogramming system, multiple processes can compete for the CPU at the same time. This situation happens whenever there are two or more processes simultaneously in the ready state. CPU scheduling plays an important role in determining which process should be moved from one state to another. There are two types of scheduling: non-preemptive scheduling【非抢占式调度】, and preemptive scheduling【抢占式调度】. Non-preemptive scheduling is used when a process switches from the running state to the waiting state or the terminated state. In this mode, a running process is executed until it exits or voluntarily blocks itself to wait for I/O or some operating system service. Preemptive scheduling is used when a process moves from running state to the ready state or from the waiting state to the ready state. In this mode, the operating system is allowed to interrupt【中断】 the currently running process and move it to the ready state. The part of the operating system that makes the choice is called the scheduler【调度器】, and the algorithm it uses is called the scheduling algorithm【调度算法】. We will introduce several algorithms in the following sections.

1. First-Come First-Served

First-come first-served (FCFS)【先来先服务】 is probably the simplest of all scheduling algorithms. It is non-preemptive. With this algorithm, processes are assigned the CPU in the order they arrive in the running state. Suppose processes p1 through p4 request CPU resources in the same time but in the following order with different service time (Table 7.1).

Table 7.1 Process and service time in FCFS

Process	Service Time
p1	100
p2	75
p3	200
p4	125

Assume processes do not cause themselves to wait. Gantt chart【甘特图】 above shows the order and time of process completion (Figure 7.11). Because we assume all processes arrived at the same time, the average turnaround【轮转时间】 time should be (100+175+375+500)/4 = 287.5.

Unfortunately, FCFS has a powerful disadvantage that it pays little attention to important factors such as service time requirements. And it does not use the different service time to determine the best order, either.

2. Shortest Job First

The shortest-job-first (SJF)【最短作业优先】 is also a non-preemptive algorithm. When several equally important processes are requesting the CPU resources, the scheduler always picks up the one with smallest service time. Below is the Gantt chart for the same set of processes in the FCFS example (Figure 7.12).

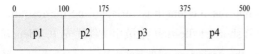

Figure 7.11 Example of FCFS scheduling

Figure 7.12 Example of SJF scheduling

Similarly, we assume all processes arrived at the same time and have no waiting time, the average turnaround time should be (75+175+300+500)/4 = 262.5.

To run this algorithm, the operating system has to estimate【评估】 the service time【服务时间】 of all the processes. However, this algorithm is provably ideal that we cannot know the service time of each job in advance.

3. Round-Robin Scheduling

One of the oldest, simplest, fairest, and most widely used algorithms is called round-robin (RR) scheduling【轮询调度】. It distributes the processing time equally among all processes. Each process is assigned a time interval called time quantum【时间段】 or time-slice【时间片】, during which it is allowed to run. If the process used up the time-slice it is assigned, the CPU would be preempted and given to another process. This procedure continues until the process completes and terminates. It is a preemptive algorithm.

Suppose the time-slice used for round-robin algorithm was 50. Below is the Gantt chart for the same set of processes in the FCFS example (Figure 7.13).

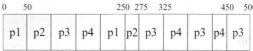

Figure 7.13 Example of round-robin scheduling

Each process gets a time quantum of 50, unless it do not need a full time-slice. For instance, process p2 requires 75 time units. When it is assigned CPU for the first time, it uses time-slice of 50. But when it is assigned CPU again in next turn, it uses only 25 time units. Therefore, it terminates at 275.

The average turnaround time should be (250+275+500+450)/4 = 368.75. This turnaround time is larger than the turnaround time in examples of algorithms mentioned above. But we cannot generally say that this algorithm is worse than another based on one specific set of processes. Generally speaking, this algorithm is considered the most fair.

4. Other Scheduling algorithms

There are even more scheduling algorithms. Here are some examples.

- Priority scheduling【优先级调度】: Each process is assigned a priority, and CPU is allocated to the runnable process with the highest priority. Equal priority processes are scheduled in FCFS order.
- Multilevel Queue【多级队列】 scheduling: Set up the priority queues【优先级队列】 and put different classes of priority in each queue. Each queue implements whatever scheduling algorithm is most suitable to its processes. Processes in the highest priority queue were run (possibly in a RR manner). Whenever a process used up all the time-slice allocated to it, it is moved to a lower priority queue. Whenever a process does not use the full time-slice (e.g. in an interactive process), it is moved to a higher priority queue.

7.2.4 Device manager, file manager, and user interface

Device manager is responsible for managing the access to input/output devices. File manager is responsible for controlling the access to files. The user interface is responsible for communication outside the operating system (like a public relations department, see Figure 7.3). As they are beyond the scope of this book, we will not discuss the details.

7.3 Popular Operating Systems

7.3.1 UNIX

UNIX, developed in 1969 by a group of AT&T employees at Bell Labs, is a family of portable, multitasking【多任务】, multiuser【多用户】, and time-sharing operating system. It is a very powerful operating system with several excellent features:

- It is written mostly in C language 【C语言】 which makes it an easily readable【可读的】, modifiable【可修改的】, and portable【可移植的】 operating system.
- It provides abundant, well-chosen system calls【系统调用】 that can solve many problems efficiently.
- It provides a powerful and programmable shell language as user interface that makes the system easy to use.
- It can be easily configured to run devices because of the integration of the device drivers【设备驱动】 in operating system.

7.3.2 Linux

Linux evolved from a kernel developed by Linus Torvalds, a Finnish student at the University of Helsinki, in 1991. It is based on UNIX. Linux was originally developed for personal computers based on the Intel x86 architecture【架构】, but has since been portable to more computer hardware platforms than any other operating system. It is a free【免费的】 operating system for every user while UNIX is not free. The initial Linux kernel, which was similar to a small subset of UNIX, has grown into a full-fledged 【成熟的】 operating system today.

7.3.3 Windows NT series

In the late 1980s Microsoft, under the leadership of Dave Cutler, started development of a new single-user operating system to replace MS-DOS (Microsoft Disk Operating System). Windows NT was the outcome. Windows NT Workstation 4.0, Windows 2000 Professional, and Windows XP fall into Windows NT family. NT stands for "New Technology" but no longer carries any particular meaning.

Well-known operating systems, Windows 7, Windows 8, and Windows 10 also fall into Windows NT family. Windows 7 was released to manufacturers on July 22, 2009, which was an incremental upgrade【增量式升级】 from Windows Vista. To compete with other mobile operating systems such as Android and iOS, Microsoft officially released Windows 8 on August 1, 2012. It improves its user experience【用户体验】 on tablets. Although it has improved performance, enhanced security, and improved support for touchscreen devices, it was widely criticized for being difficult to learn. In September 2014, Microsoft officially unveiled Windows 10 and released to consumers on July 29, 2015. It introduces so-called "universal apps", that can run across Microsoft product families with almost same code. It also revised part of the user interface【用户界面】 that was widely criticized since Windows 8.

7.3.4 Windows Phone

Windows Phone is a mobile operating system first released in 2010, by Microsoft. It supports a variety of mobile devices, such as tablets, and smartphones. The first released version was named as Windows Phone 7.0. In 2012, Microsoft released a brand new system called Windows Phone 8.0, which uses the kernel【内核】 of Windows NT.

7.3.5 iOS

iOS is derived from Mac OS X, with which it shares the Darwin foundation, and is therefore a UNIX-like operating system by nature. iOS is Apple's mobile version of the Mac OS X operating system used on Apple computers. Apple does not license【许可】 iOS for installation on non-Apple hardware. When it comes to device sales, after Android, it is the second most dominant mobile operating system platform in the world.

7.3.6 Android

Android is a Linux-based operating system designed primarily for touch screen mobile devices【触屏移动设备】 such as smart phones and tablets. It was initially developed by Android Inc., and financially backed and purchased in 2005 by Google. It is an open-source【开源的】 operating system, and can be used in all kinds of mobile devices. Like iOS, Android's user interface is based on direct manipulation and also has a good user experience (good response, all kinds of applications).

7.3.7 Chrome OS

Chrome OS is also a Linux-based operating system developed by Google in 2007, and makes it an open-source project in 2009. It is conceived as a cloud-centric operating system【云操作系统】, hence it is closely integrated with Internet【互联网】. In this OS, both applications and user data reside in the cloud【云端】. The first laptop that installed Chrome OS arrived in May 2011, known as Chromebook.

7.3.8 Comparison of operating systems

This section introduces some popular operating systems. Let us make a brief generalization and comparison of these operating systems here (Table 7.2).

Table 7.2 Comparison of operating systems

Name	Target system type	OS Family	Advantages	Disadvantages
UNIX	Server, Personal computer, Embedded devices	UNIX	1. With compact file system. 2. Portable【可移植的】 and stable【稳定的】	Traditional UNIX kernel is inflexible and lack extensibility【可扩展性】
Linux	Server, Personal computer, Embedded devices	Linux	1. Open-source【开源的】 and customizable【可定制的】. 2. Secure, stable and easy to maintain【维护】. 3. Most programs are free and open-source	1. Hard to use for most users. 2. Lack of applications such as games

续表

Name	Target system type	OS Family	Advantages	Disadvantages
Windows (NT family)	Workstation, Personal computer, Embedded devices, Tablet PC	Windows NT	1. With a user-friendly GUI (Graphical User Interface)【用户图形界面】, and simple to use. 2. Provide a high performance platform for variable applications	With relatively more bugs and cannot provide system updates in time
Mac OS	Workstation, Personal computer, Embedded devices	Linux	1. Easy to install and use. 2. Secure and occupy less system resources	Hard to customize the operating system
Chrome OS	Chromebook	Linux	1. All applications and user data reside in the cloud. 2. Fast, secure and easy to use	1. Lack of applications such as games. 2. Performance of operating system is not powerful
iOS	Smartphone, Tablet	Darwin (UNIX)	1. With a user-friendly GUI, and simple to use. 2. Secure, stable, and smooth【流畅的】	1. Close and hard to customize. 2. Can only run on Apple's devices
Android	Smartphone, Tablet	Linux	1. Open-source, free, and customizable. 2. Portable and extensible	1. Less restrictions may cause security problems. 2. Personal privacy is easy to be exposed
Windows Phone	Smartphone	Windows NT	1. Easy integration with common Microsoft programs. 2. Distinctive GUI. 3. Smooth and secure	1. Small number of applications available. 2. Unfriendly support for system update

7.3.9 32-bit operating systems vs 64-bit operating systems

After the discussion of modern popular operating systems, let us talk about an important concept in operating system that has nothing to do with specific operating systems. That is **32-bit operating system**【32位操作系统】, and **64-bit operating system**【64位操作系统】. Here, 32-bit or 64-bit means the size of CPU and ALU (Arithmetic Logical Unit)【算术逻辑单元】architectures are based on registers【寄存器】, address buses【地址总线】, or data buses【数据总线】of 32-bit or 64-bit respectively.

Registers are designed to store the address in the physical or virtual memory of the computer, therefore, the total number of addresses is often decided by the width of registers. A 32-bit address register meant that 2^{32} addresses, or 4 GB of RAM could be referenced. However, 32 bit operating system was common until the early 1990s, when the continual decrease in the cost【成本】of memory led to installations with quantities of RAM approaching 4 GB, and the use of virtual memory spaces exceeding the 4 GB became desirable. To solve this problem, 64-bit microprocessor architectures【微处理器架构】were developed, which means the register can access 2^{64} addresses or 16EB of RAM. Furthermore, a 128 bit wide ALU means there have to be 128 bit wide address buses or data buses, and registers, cache, other memory, etc. would be needed to effectively use the wide ALU. And in terms of cost, the wide ALU doesn't only make it expensive, but other parts of the chip as well. So 64-bit operating system is quite enough for modern computers.

The following is a brief comparison of 32-bit operating system and 64-bit operating system (Table 7.3).

Table 7.3 32-bit operating system vs 64-bit operating system

Description	32-bit	64-bit
Version of runnable application【可运行应用】	32-bit application	32-bit application, 64-bit application
Maximum usable memory	2^{32} bytes	2^{64} bytes
Size of registers	2^{32} bit	2^{64} bit
Advantages	Less memory usage	1. Larger usable memory. 2. Better compatibility. 3. Better performance for running applications
Disadvantages	Maximum usable memory is 2^{32} bit	More memory usage than 32-bit application, when running the same application

7.4 References and Recommended Readings

For more details about the subjects discussed in this chapter, the following are recommended:
- Abraham S, Galvin P B, Gagne G. Operating System Concepts. Wiley, 2012.
- Hermann K, Real-Time Systems. Design Principles for Distributed Embedded Applications. Springer, 2011.
- Love R. Linux Kernel Development. Pearson, 2010.
- McHoes A, Flynn I M. Understanding Operating Systems. Boston, MA: Cengage Learning, 2014.
- Patterson D A, Hennessy J L. Computer Organization and Design: The Hardware/Software Interface. Waltham, MA: Elsevier, 2014.
- Stallings W. Operating Systems: Internals and Design Principles. Pearson, 2014.
- Tanenbaum A S, Bos H. Modern Operating Systems. Pearson, 2014.
- Tanenbaum A S, Woodhull A S. Operating System Design and Implementation. Upper Saddle River, NJ: Pearson, 2006.
- http://www.apple.com/cn/ios/what-is/, Retrieved on April, 2016.
- http://developer.android.com/index.html, Retrieved on April, 2016.

7.5 Summary

- An operating system provides common services to other programs, and ensures the efficient use of hardware and software resources.
- The evolution progress of operating system includes, but not limited to, batch operating system, time-sharing systems, single-user operating system, parallel systems, distributed systems, and real-time systems.
- The basic components of the operating system are: memory manager, process manager, device manager, file manager, and user interface.

- Address space is a set of addresses that a program can use to address memory.
- Multiprogramming allows multiple programs to be performed at the same time, while monoprogramming allows only one program.
- Virtual memory means each program has its own address space, which is broken up into small units called pages.
- Partitioning means dividing the main memory into multiple partitions, usually with contiguous and variable length of address spaces.
- In paging, memory is divided into fixed-size units called frames and the program is divided into pages of the same size. There is no need for pages of program to be contiguous, but all pages must be present in memory for execution.
- By using swapping technique, demand paging has removed the restriction that the entire program needs to be in memory.
- Segmentation is a technique that partition memory into logically related units with variable length, called segments, which is implemented by segment table. Demand paging and segmentation can be combined in order to improve the efficiency of the computer system.
- The process schedulers moves a process from one state to another. Jobs and processes wait in queues.
- There are two types of scheduling: non-preemptive scheduling, and preemptive scheduling.
- First-come first-served assigns the CPU resources according to the arriving order.
- Shortest-job-first assigns the CPU resources to the jobs with short service time.
- Round-robin scheduling rotates the CPU resources among active processes, giving little bursts of computation to each process.
- There are other scheduling algorithms such as: priority scheduling, and multi-level queue scheduling.
- Deadlock is a situation in which a process cannot execute because of the unrestricted use of resources.
- Starvation is a situation in which a process cannot execute because of too many restrictions on resources.
- UNIX, Linux, Windows NT Series, iOS, Android, Windows Phone, and Chrome OS are some of the popular operating systems.
- 32-bit and 64-bit in operating systems mean the size of CPU and ALU architectures that are based on registers, address buses, or data buses of 32-bit or 64-bit respectively.
- A 32-bit address register meant that 2^{32} addresses of RAM could be referenced. A 64-bit address register meant that 2^{64} addresses of RAM could be referenced.

7.6 Practice Set

1. State the differences between monoprogramming and multiprogramming.
2. What is partitioning?

3. How is demand paging more efficient than regular paging?
4. Describe the differences between paging and segmentation.
5. Describe the difference between virtual memory and physical memory.
6. Describe all the states a process can be in.
7. If a process is in the ready state, what states can it go next?
8. Describe the difference between a job scheduler and a process scheduler.
9. How is deadlock different from starvation?
10. List some currently popular operating systems.
11. Multiprogramming requires a _____ operating system.
    a. batch	b. time-sharing	c. real-time	d. parallel
12. _____ use the technique of swapping.
    a. Partitioning	b. Virtual memory
    c. Demand paging	d. Synchronization
13. _____ is a multiprogramming technique in which multiple programs reside entirely in memory with each program occupying consecutive memory locations.
    a. Partitioning	b. Paging	c. Demand paging	d. Segmentation
14. In _____, the program can be divided into variable-length sections.
    a. Partitioning	b. Virtual memory
    c. Demand paging	d. Queuing
15. A process in the _____ state can go to either the ready, terminated, or waiting state.
    a. hold	b. virtual	c. running	d. a and c
16. A program becomes a _____ when it is brought to the hold state.
    a. job	b. process	c. program	d. partition
17. The _____ scheduler can change from a job or vice versa.
    a. job	b. process	c. virtual	d. queue
18. To prevent _____, an operating system can make resource restrictions on processes.
    a. starvation	b. queue	c. synchronization	d. deadlock
19. The _____ manager control the access to I/O devices.
    a. memory	b. process	c. device	d. file
20. A computer has a monoprogramming operating system. If there are 512 MB memory available and the operating system needs 64 MB, what is the maximum size of a program that can be run?
21. Redo Exercise 20 if the operating system automatically allocates 30 MB of memory to data.
22. A monoprogramming operating system executes programs that on average require 10 microseconds access to the CPU and 90 microseconds access to the I/O devices. What percentage of time is the CPU busy?
23. A multiprogramming operating system uses paging to manage memory. The size of the memory is 512 MB and is divided into 128 pages, each page with 4 MB. The first program needs 13 MB. The second program needs 12 MB, and the third program needs 27 MB. Answer the following questions.
    a. How many pages are used by the first program?
    b. How many pages are used by the second program?

c. How many pages are used by the third program?

d. How many pages are unused?

e. What is the total memory wasted?

f. What percentage of memory is wasted?

24. Assume the operating system do not use the virtual memory but requires the whole program to be in physical memory during execution (no paging or segmentation). The size of physical memory is 200MB. The size of virtual memory is 1 GB.

a. How many programs of size 10 MB can be executed concurrently by this operating system?

b. How many of them can reside in memory at any time?

c. How many of them must be on the disk?

25. Table 7.4 shows a set of processes, assuming the operating system uses FCFS, SJF, and round-robin scheduling respectively for CPU scheduling, write down the average turnaround time of these three algorithms.

Table 7.4 Set of processes

Process	Service Time
p1	150
p2	175
p3	320
p4	25
p5	125

Unit 8 Algorithm

In Chapters 2 to 4 we have introduced number systems, and how to store and manipulate data in a computer. They are the basic information necessary to understand a computing system【计算系统】. In Chapters 5 to 7 we have introduced the hardware and software components of a computer【计算机的硬件和软件部件】. In the above chapters, the discussion has focused primarily on what a computer system is. Now, we will focus on how to use a computer system.

Algorithm【算法】is the most fundamental concept in computer science. In this part, we introduce the basic concept of algorithm and the constructs for developing algorithms【用于开发算法的结构】, then we discuss the indicators for evaluating algorithms 【用于评价算法的指标】 and list the basic algorithms【基本的算法】. Finally, we introduce the classification of algorithms【算法的分类】.

8.1 What Is Algorithm?

8.1.1 Problem solving

Problem solving【问题求解】 is to find a solution for a complicated, distressing, vexing, or unsettled question. Computers are not intelligent【智能的】, they cannot analyze problems and propose a solution. If they are not told what to do, they can do nothing. So, a person (the programmer【程序员】) must analyze the problem first, and then create the instructions【指令】(the program【程序】), finally, the computers carry out the instructions.

Since computers cannot solve certain problems, why do we still use it? In fact, once we have made a solution (an instruction sequence) for the computer, the computer can repeat the instruction sequence very quickly and consistently for different situations and data【对于不同的数据环境，计算机能够快速而一致的重复执行这些指令序列】. In this way, we can be liberated from repetitive and dull jobs. In computer science, we call the plan of solution an algorithm.

8.1.2 Definition of algorithm

An algorithm is a series of clear instructions to solve the problem, that is, you can get the required output【要求的输出】 in a limited time for a normative input【规范的输入】.

Algorithms have five basic characteristics: input 【输入】, output【输出】, finiteness【有穷性】,

definiteness【确定性】, and effectiveness【可行性】.

- **Input:** An algorithm must have zero or multiple inputs that inscribe the initial condition of the operands【刻画运算对象的初始情况】.
- **Output:** An algorithm must have one or more outputs to reflect the results of input data processing【反映输入数据的处理结果】.
- **Finiteness:** An algorithm must be able to terminate after executing finite steps (halt)【执行有限步骤后终止（中止）】.
- **Definiteness**: Each step of the algorithm must have the exact definition.
- **Effectiveness:** Each step in the algorithm should be able to be effectively implement, and get certain result.

8.2 Three Constructs

In recent years, structured programming thought【结构化程序设计思想】 is becoming increasingly enjoying popular support. With structured programming, you can vastly speed up the application development【应用程序开发】, and improve the readability of the code【代码可读性】. Structured programming is made up of a combination of several basic structures. There are three constructs for a structured program or algorithm: sequence【顺序结构】, decision (selection)【选择结构】, and repetition【循环结构】 (Figure 8.1). And it has been shown that other structures are no longer required. Besides the designing difficulty will be decreased, the readability and the maintainability【可维护性】 will be increased if these constructs are applied to compilers.

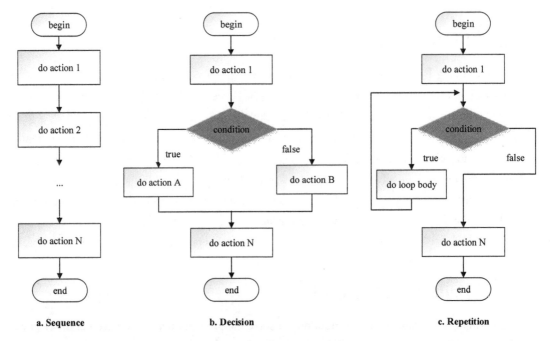

Figure 8.1   Three constructs

Sequence: The simplest, and the most common program structure. It is a sequence of instructions, which can be in top-down order【自上而下，依次执行】. But in most cases, it is used as a part of the program. It constitutes a complex program with the other two constructs.

Decision: It is used when you need to make a judgment. You can decide which sequence of instructions to follow according to the result of judgment.

Repetition: In some problems, we should repeat the same sequence of instructions. In that case, we can use a construct of this kind.

8.3　How to Evaluate Algorithms

Algorithm is a solution to the problem, but there may be several solutions for a problem. Naturally, we need to make a comparison of advantages and disadvantages among these different solutions. For example, when there is only Dave to clean windows, you cannot find if his work is good or not, but when there are Dave and Mary or even more people to do the same work, then we could assess the performance according to the different evaluation standard. Some think Dave is the best, because he is the fastest, while others believe that Mary is the best, because she did a good job of wiping clean.

Algorithm is the kernel of programming, the quality of an algorithm significantly affects the performance of application【算法是程序设计的核心，算法的质量严重影响应用程序的性能】. So how to evaluate algorithms? We have the following several evaluation standards:

1. Correctness【正确性】

An algorithm has a significance of existence only if it is correct. It is the most important performance index. So, make sure that programmers use the correct computer programming language【计算机编程语言】 to achieve the algorithm.

2. Readability【可读性】

It refers to the convenience of reading the algorithm. The algorithm may be implemented after many revisions, or it can be ported【被移植】 to other functions. Thus, algorithms should be readable, and easier to understand. In this way, it is convenient for programmers to analyze and modify the algorithm, and translate it to other functions【方便程序人员对其进行分析、修改，并将其移植到其他函数中】.

3. Robustness【健壮性】

It refers to the response and processing ability of the unreasonable data input, also called fault-tolerance【容错性】.

4. The time complexity【时间复杂度】

It refers to the computer time required to execute the algorithm. The time complexity of an algorithm is commonly expressed using O notation, which excludes coefficients and lower order terms. When expressed this way, the time complexity is said to be described asymptotically, i.e., as the input size goes to infinity. Generally, algorithms grows as a function (such as $f(n)$) of input size n【算法是问

题规模 n 的函数 f(n)】，therefore, the time complexity of algorithm is marked as: $T(n)=O(f(n))$. For example, if the time required by an algorithm on all inputs of size n is at most $5n^3+3n$ for any n (bigger than some n_0), the asymptotic time complexity is $O(n^3)$.

5. The space complexity【空间复杂度】

It refers to the consumed memory space of the algorithm. Its calculation and representation is similar to the time complexity.

8.4 Algorithm Representation

An algorithm is a sequence of instructions or a set of rules that are carried out to perform a task. This task can be anything, so long as you can give clear instructions for it. There are many kinds of notation to describe algorithms, such as natural languages【自然语言】, pseudocode【伪代码】, UML (Unified Modeling Language) diagrams【统一建模语言图】, and programming languages【编程语言】. Natural language expressions of algorithms is easy to understand, but they are often lengthy or inherently ambiguous【算法的自然语言表达容易被理解，但它们往往是冗长或有歧义的】. So, they have rarely been applied to complex or technical algorithms. Pseudocode and UML are structured ways to express algorithms that avoid many of the ambiguities common in natural language statements. Programming languages are primarily intended for expressing algorithms in a form that can be executed by a computer, but are often used as a way to define document or algorithms. Here we introduce UML and pseudocode in detail.

8.4.1 UML

Unified Modeling Language (UML) uses the big structure flowcharts【结构化流程图】 to show how the algorithm flows from beginning to end. It is a graphical representation of an algorithm, and hides all details of the algorithm (Figure 8.2).

Problem: Write an algorithm to find the result of (a+b)∗c−d.

Solution: Use the sequence construct (Figure 8.1).

8.4.2 Pseudocode

Pseudocode is something in between natural languages and programming languages. It is a kind of informal, English-like representation of an algorithm, and is used to describe module structure diagram【模块结构图】. Using pseudocode, the algorithm can be implemented in any kind of programming language (Pascal, C, Java, etc.) easily.

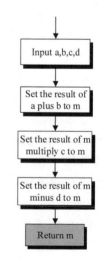

Figure 8.2　UML for (a+b)∗c−d

Problem: Write an algorithm to find the largest of 100 numbers.

Solution: Use the repetition construct and decision construct (Algorithm 8.1).

Algorithm 8.1　Find the largest of 100 numbers

Input: 100 positive integers list
Output: the largest of 100 numbers
Algorithm FindLargest(list)
{
1. Set Largest to $-\infty$
2. Set Counter to 0
3. while (Counter less than 100)
　　3.1　if (the integer is greater than Largest)
　　　　　then
　　　　　　　3.1.1　Set Largest to the value of the integer
　　　　End if
　　3.2　Increment Counter
　End while
4. Return Largest
}

8.5　Basic Algorithms

In computer science, several basic algorithms are used frequently, such as summation【求和】, product【乘积】, largest algorithms【最大值算法】, sorting【排序】algorithms, and searching algorithms【搜索算法；查找算法】. We will focus on sorting and searching algorithms here.

8.5.1　Sorting Algorithms

Sorting algorithms put elements into ordered arrays【有序数组】. Sorting a list of items in ascending or descending order【升序或降序】 can help either a human or a computer find items in that list quickly. In this section, we introduce a few fairly different sorting algorithms, and then you'll find that there are many different ways to solve the same problem.

1. Selection sort

The selection sort【选择排序】algorithm is a simple and intuitive sorting algorithm. It reflects how we can sort a list of values by hand. The basic steps are as follows:

（1）Select the minimum element from the unsorted list【未排序列表】.
（2）Put it at the end of the sorted list sequentially【依次将它放在已排序列表的表尾】.
（3）Repeat until all elements are ordered.

The translation of this algorithm into a computer program is simple, but it has one drawback: It requires space for two complete lists【它需要两个完整的列表空间】. However, we can make a minor adjustment to get rid of the need to this duplicate space. The solution is to exchange the minimum element with the first element in the unsorted list, and the new unsorted list starting with the second element in previous unsorted list (Figure 8.3).

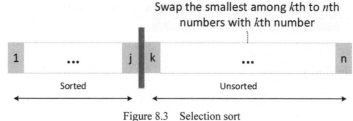

Figure 8.3 Selection sort

Let's look at an example (Figure 8.4).

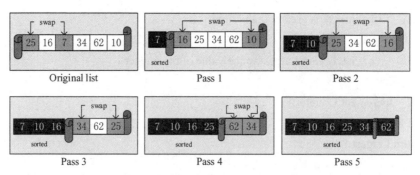

Figure 8.4 Example of selection sort

Clearly, the lists are divided into two parts: the sorted part (shaded), and the unsorted part (not shaded). At first, all data elements are put in the unsorted part, then you move one element from the unsorted part to the sorted part, so you have completed a sort pass. Finally, all data elements are put in the sorted part. The implementations of selection sort could produce an unstable sort【不稳定排序】. A sorting algorithm is said to be stable if two elements with equal keys (i.e. values) retain the relative order of values in the sorted output as they were in the input unsorted array or list. Figure 8.5 shows the UML for the selection sort algorithm.

2. Bubble sort

The bubble sort【冒泡排序】 is the simplest sorting method. The principle of this method is to regard the unsorted elements as mixed sizes for bubbles【此方法的基本思想是将无序元素看作大小混合的气泡】. The smaller elements are comparatively light, thus, they will be floating upward. Start from the last element in the list,

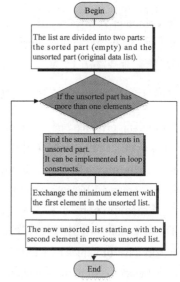

Figure 8.5 Flowchart for selection sort algorithm

and compare the adjacent element pairs【比较相邻元素对】, when the bottom element of the pair is smaller than the top, swap them (Figure 8.6). When accessed to the whole unsorted list data, the smallest bubble will be rising to the top. Repeat until no swaps are needed and all elements are ordered. The implementation of bubble sort could produce a stable sort【稳定排序】.

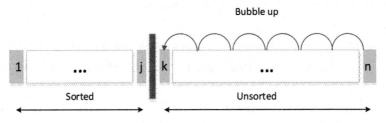

Figure 8.6   Bubble sort

Let's look at an example (Figure 8.7).

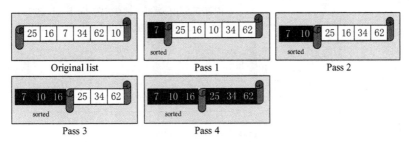

Figure 8.7   Example of bubble sort

3. Merge sort

Merge operation【归并操作】is the method of merging two ordered sequences into one. Merge sort【归并排序】is an efficient sorting algorithm based on merging operations. The core idea of this method is to regard a list with N elements as N sorted sublists【子列表】, each sublist contains one element (a list of one element is considered sorted). And then merge every two sublists to produce N/2 new sorted sublists. Repeatedly merge sublists until there is only one sublist remaining. This will be the sorted list (Figure 8.8). Most implementations of merge sort produce a stable sort.

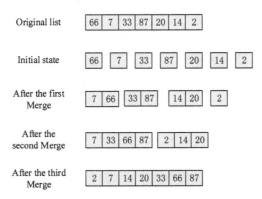

Figure 8.8   Example of merge sort

4. Insertion sort

The insertion sort【插入排序】 algorithm is one of the most common sorting techniques, and it is often unknowingly used by card players when sorting the cards in their hands into order. Each card a player picks up is inserted into its proper position in their hand of cards to maintain a specific order. Figure 8.9 shows the basic steps as follows:

(1) Regard the first element of original list as a sorted list, and regard the second to last element of original list as an unsorted list.

(2) Insert the first element of unsorted list at appropriate place in sorted list. (If the inserted element is equal to one of the elements in the sorted sequence, then put the inserted element at the end of the equal element of sorted list).

(3) Repeat until all elements are ordered.

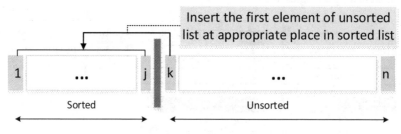

Figure 8.9　Insertion sort

Let's look at an example (Figure 8.10); the original list is 25, 16, 7, 34, 62, 10.

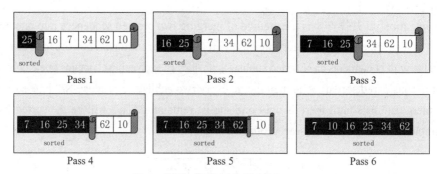

Figure 8.10　Example of insertion sort

5. Shell sort

Shell sort (also known as Diminishing Increment Sort)【希尔排序（缩小增量排序）】 is a more efficient improved version of insertion sort. The basic steps are as follows:

(1) Take an integer d_1 that is less than n (n is the number of elements in original list).

(2) When the distances of two elements equal d_1 or time it, they will be divided into a group. And all elements of unsorted list are divided into d_1 groups.

(3) Each group can be sorted by an insertion sort.

(4) Take an integer $d_2<d_1$, repeat the previous step, until $d_t=1$ ($d_t < d_{t-1} < \cdots < d_2 < d_1$).

Let's look at an example (Figure 8.11), the original list is 6, 18, 23, 15, 32, 8, 78, 17, 48, 65, 5, 27.

6. Quick sort

Quick sort【快速排序】 is a divide and conquer algorithm【分治算法】. Quick sort first divides a large array【数组】 into two smaller sub-arrays【子数组】: the low elements and the high elements (Figure 8.12). Quick sort can then recursively【递归地】 sort the sub-arrays. The basic steps are as follows:

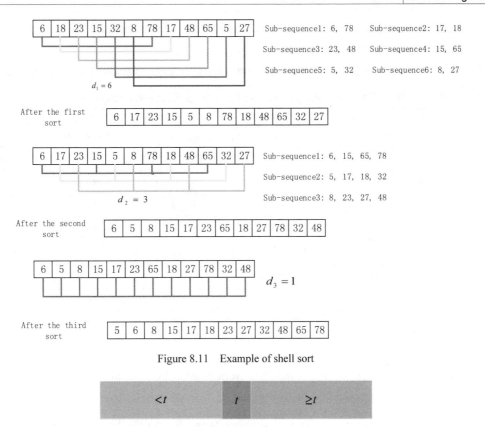

Figure 8.11 Example of shell sort

Figure 8.12 Quick sort

（1）Pick an element, called a pivot【基准】, from the array.

（2）Partitioning: Reorder the array so that all elements with values less than the pivot come before the pivot, while all elements with values greater than the pivot come after it (equal values can go either way). After this partitioning, the pivot is in its final position. This is referred to as the partition operation.

（3）Recursively apply the above steps to the sub-array of elements with smaller values and separately to the sub-array of elements with greater values.

The base case of the recursion is arrays of size zero or one, which never need to be sorted. (Base case does not make a recursive call.) There are several different strategies for pivot selection and partitioning; the performance of quick sort is greatly affected by the choice of specific implementation schemes.

Let's look at an example (Figure 8.13).

7. Heap sort

The heap sort【堆排序】 is a sorting algorithm designed by using heap data structure【堆数据结构】. The algorithm involves preparing the list by first turning it

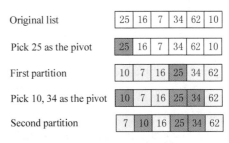

Figure 8.13 Example of Quick sort

into a max heap, then repeatedly swaps the first value of the list with the last value, decreasing the range of values considered in the heap operation by one, and sifting the new first value into its position in the heap. This repeats until the range of considered values is one value in length. The implementations of heap sort could produce an unstable sort.

Finally, we give the simple comparison of several basic sorting algorithms index (Table 8.1).

Table 8.1 Complexity comparison of several basic sorting algorithms

Sorting Method	Time Complexity			Space Complexity Worst Case	Stability
	Best Case	Average Case	Worst Case		
Selection sort	$O(n^2)$	$O(n^2)$	$O(n^2)$	$O(1)$	Unstable
Bubble sort	$O(n)$	$O(n^2)$	$O(n^2)$	$O(1)$	Stable
Merge sort	$O(n\log(n))$	$O(n\log(n))$	$O(n\log(n))$	$O(n)$	Stable
Insertion sort	$O(n)$	$O(n^2)$	$O(n^2)$	$O(1)$	Stable
Shell sort*	$O(n\log(n))$	$O(n\log^2(n))$	$O(n\log^2(n))$	$O(1)$	Unstable
Quick sort	$O(n\log(n))$	$O(n\log(n))$	$O(n^2)$	$O(n)$	Unstable
Heap sort	$O(n\log(n))$	$O(n\log(n))$	$O(n\log(n))$	$O(1)$	Unstable

* The exact complexity of shell sort algorithm varies depending on the increment sequence (or gap sequence) of a particular implementation. For more details, please see (Weiss M A, Data Structures and Algorithm Analysis in C++, New Jersey, Addison Wesley 2014).

Complexity rating table of big-O notation from fastest (best) to slowest (worst) is shown in Table 8.2.

Table 8.2 Big-O complexity rating table【复杂性程度分级表】

Rating【程度分级】	Big-O complexity【复杂性表达式】	
Best (fastest)	Logarithmic【对数】: $O(\log(n))$	Constant【常数】: $O(1)$
Good	Linear【线性】: $O(n)$	
Fair	Linearithmic【对数线性】: $O(n\log(n))$	Polylogarithmic【多项式对数】: $O(n\log^2(n))$
Bad	Quadratic【二次方】: $O(n^2)$	Exponential【指数】: $O(2^n)$
Worst (slowest)	Factorial【阶乘】: $O(n!)$	

8.5.2 Searching algorithms

Searching【搜索；查找】 is the work of finding a particular data element in a large number of data elements. In the case of a data list, each data element is identified by keyword【关键字】. Searching means that according to a given value, you need to find the location (index)【位置（或索引）】 of the first element in the list that is equal to that value. There are several kinds of searching algorithms, such as sequential search【顺序查找】, binary search【二分查找】, block search【分块查找】, and hash table search【哈希表查找】. Here we introduce sequential search and binary search in detail:

- Sequential search: It can be used for unordered (or ordered) lists.
- Binary search: It can only be used in an ordered list.

1. Sequential search

Sequential search is the simplest search method for a disordered list. When there are a few elements in the list, or lists are not searched frequently, you can use this method. Otherwise, it is better to sort the list first, and then search it using the binary search method discussed later.

The principle of sequential search is to make a comparison between the target and the list element one by one from the first element of the list until you either find the target or prove that it is not in the list【顺序查找的基本原理是让目标与列表中的元素从第一个开始逐个比较，直到找到目标或证明其不在列表中】. Figure 8.14 shows the steps to search the value 4.

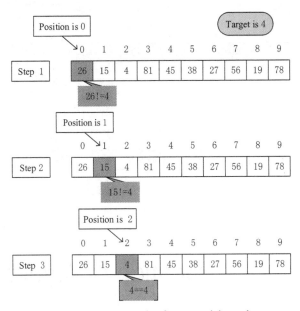

Figure 8.14　Example of a sequential search

2. Binary search

Obviously, the sequential search algorithm is slow and inefficient than the binary search algorithm. If we have a list of a million or more elements, we must make a million or more comparisons in the worst case. If the list you got is not ordered, the sequential search method will be the only choice. But when the list is ordered, you had better use a more efficient algorithm called binary search.

The basic steps are as follows:

（1）Determine the element at the middle of the list: mid=(low+high)/2.

(mid represents the location of the element in the middle of the list, low represents the leftmost one, and high represents the rightmost one.)

（2）Compare the target with the keyword at the middle of the list. The search is successful if the two are equal, otherwise you are required to determine the new search range【确定新的查找范围】:

a) if the target is greater than the mid, then low= mid+1;

b) if the target is smaller than the mid, then low= mid-1;

c) if low=high, we can conclude that the target is not in the list.

（3）Repeat this process until you either find the target or prove that it is not in the list.

Figure 8.15 shows how to find the target 45.

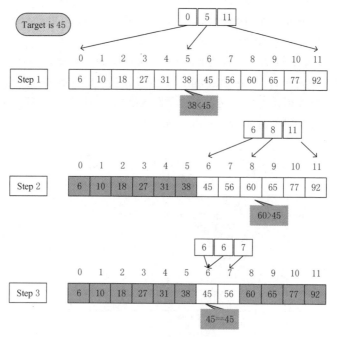

Figure 8.15　Example of a binary search

8.6　Classification of Algorithm

8.6.1　By implementation

An algorithm may be implemented by different basic principles.

1. Recursion or iteration

A recursion【递归】 algorithm is a special process which calls itself directly or indirectly【直接或间接地调用自身】 to solve a problem. Iteration【迭代】 algorithms use repetitive constructs like loops to execute a series of instructions until the termination condition is reached. In theory, every recursive program could be replaced by an equivalent (but possibly more or less complex) iterative version, and vice versa (Algorithm 8.2 and 8.3).

Algorithm 8.2　Iterative factorial

Input: A positive integer num
Output: Factorial of the num **FN**
Algorithm Factorial(num)
{
1. Set FN to 1
2. Set i to 1
3. while (i is less than or equal to num)
　　3.1　Set FN to FN × i

>   3.2   Increment i
>   End while
>   4. Return FN
> }

Algorithm 8.3   Recursive factorial

> Input: A positive integer num
> Output: Factorial of the num
> Algorithm Factorial(num)
> {
> 1. if (num is equal to 0)
>   then
>   1.1   return 1
> 2. else
>   2.1   return num × Factorial (num – 1)
>   End if
> }

2. Serial or parallel or distributed

Usually, we assume that computers execute just one instruction of an algorithm at a time【通常，我们假定计算机在同一时间只能执行一条算法指令】. We call them serial computers【串行计算机】. An algorithm designed for such an environment is called a serial algorithm【串行算法】. Parallel algorithms【并行算法】 are the methods and procedures of solving problems on a multi-processor parallel computer【多处理器并行计算机】, whereas distributed algorithms【分布式算法】 use multiple network-connected computers【多个通过网络连接的计算机】 to solve the same problems. Distribution【分布性】 and concurrency【并发性】 are two basic characteristics of distributed algorithm. Parallel or distributed algorithms divide the problem into more symmetrical or asymmetrical sub-problems【对称/非对称子问题】, and then collect the results of each sub-task【子任务】 back together.

Some problems are easy to be cut apart into pieces like this. For example, when you want to test all of the numbers from one to a million to see which are primes【素数】, you can divide the sets of numbers into several subsets【子集】, and assign a subset of the numbers to each available processor【处理器】, and finally, put the list of positive results back together. However, some problems cannot be easily split up into parallel portions since they need the results from a preceding step to effectively carry on with the next step【然而，一些问题不能很容易地被分割成平行部分，因为它们需要通过前一步的结果，有效地进行下一步】. We call such problems as inherently serial problems【我们称这种问题为固有串行问题】. For instance, most of the several available algorithms to compute pi (π).

3. Deterministic or non-deterministic

Given a certain set of input, a deterministic algorithm【确定性算法】 is an algorithm which will always give the same results on different runs and whose behavior can be completely predicted, with the underlying machine always passing through the same sequence of states【确定性算法在不同的运行中总是给出同样的结果，其行为是完全可以预测的，底层机器总是通过相同的状态序列】. Deterministic algorithms are widely researched and put into application nowadays, since they can be run on real machines efficiently【因为它们可以有效地在真实的机器上运行】.

Unlike deterministic algorithm, the non-deterministic algorithm【非确定性算法】 is an algorithm that can show different results on different runs. A non-deterministic algorithm can be used when the problem solved by the algorithm inherently allows multiple outcomes (or when there is a single outcome with multiple paths by which the outcome may be discovered, each equally preferable)【本质上允许有多个输出（或当有一个单一的结果和多个结果可被发现的路径，每一个同样可取）】. Importantly, every choice the non-deterministic algorithm makes while running leads to success. A concurrent algorithm【并发算法】 can perform differently on different runs due to a race condition【由于竞争条件，并发算法会在不同的运行中表现不同】.

4. Exact or approximate

We can obtain an exact solution【精确解】 of many problems, but if the problem is NP-hard problem, it can be solved in non-polynomial time【非多项式时间（复杂度）】. In this case, we could use approximation algorithms【近似算法】 for NP-hard problems to seek an approximation solution【近似解】 that is close to the true solution.

8.6.2 By design paradigm

The design methodology or paradigm【设计方法论/范例】 of algorithms is another way to classify algorithms. There is a set number of paradigms, each are different from the others. Furthermore, every paradigm will include many different kinds of algorithms.

1. Brute-force or exhaustive search

Brute-force or exhaustive search algorithms【蛮力搜索法/穷举搜索法】 take the naive method to find the best by trying every possible solution. For example, when you want to find the divisors【约数】 of a natural number n by a brute-force algorithm, you would enumerate【列举】 all integers from 1 to the square-root of n【n 的平方根】, and test whether each of them divides n without remainder【余数】.

2. Divide and conquer

The basic idea of divide and conquer algorithms【分治算法】 is to break the problem into one or more smaller subproblems【子问题】 of the same problem (usually recursively). These subproblems are independent of each other and small enough to be solved easily. Merge sort is one example of divide and conquer. A simpler variant of divide and conquer is called a decrease and conquer algorithm【减治法】, that solves an identical subproblem and uses the solution of this subproblem to solve the bigger problem【分治算法的一个简单变体叫做减治法，解决一个相同的子问题，用子问题的解决方案去解决更大的问题】. Divide and conquer divides the problem into multiple subproblems and so the conquer stage will be more complex than decrease and conquer algorithms. An example of decrease and conquer algorithm is the binary search algorithm.

Algorithm 8.4   Divide and conquer algorithms

```
Input: problem I
Output: solution
Algorithm solve(I)
{
1.  n = size( I )
2.  if (n <= smallSize)
      2.1   Solution = directly Solve (I);
3.  else
    {
      3.1   Divide I to k subproblems {I_1,I_2,I_3,…,I_k}
      3.2   For i from 1 to k
              S_i = solve(I_i)
      3.3   solution = combine (S_i, …, S_k)
    }
4.  Return solution
}
```

3. Dynamic programming

In mathematics, computer science, and economics, dynamic programming【动态规划】is an effective method to solve complex problems by breaking them down into easier subproblems. It applies to the problems exhibiting the properties of overlapping subproblems【它适用于具有重叠子问题特征的问题】. When applicable, the method takes far less time than naive methods.

The basic idea of dynamic programming is pretty simple. In general, to solve a given problem, we need to solve different parts of the problem (subproblems), and then combine the solutions of the subproblems to reach an overall solution【全面解决方案】. Often, several of these subproblems are actually the same. The dynamic programming approach seeks to solve each subproblem only once, thus reducing the number of computations: once the solution to a given subproblem has been computed, it is stored or "memo-ized"【备忘录化】; the next time the same solution is needed, it is simply looked up. This approach is especially useful when the number of repeating subproblems grows exponentially as a function of input size【当重复子问题数量随输入规模指数增长时，这种方法尤其适用】. For example, Floyd–Warshall algorithm【弗洛伊德算法】, find the shortest path between multi-sources in a weighted graph【寻找给定的加权图中多源点之间最短路径】. You can find the shortest path between adjacent vertices【相邻顶点】.

In divide and conquer, the subproblems are more or less independent, whereas subproblems overlap in dynamic programming【分治算法子问题或多或少相互独立，而在动态规划中子问题重叠】. It is the principal difference between dynamic programming and divide and conquer algorithms.

4. Genetic Algorithms

Genetic algorithms (GAs) are powerful search algorithms that are very good at optimization problems for which the exact algorithms are of very low efficiency. In GAs, the initial set of possible solutions is used to generate a new set of possible solutions. Examples of problems solved by applying GAs are the knapsack problem and the travelling salesman problem.

8.7　References and Recommended Readings

For more details about the subjects discussed in this chapter, the following books, articles, and web resources are recommended:

- Aho A V, Hopcroft J E, Ullman J D. The Design and Analysis of Computer Algorithms. Boston, MA: Addison Wesley, 1974.
- Cormen T H, Leiserson C E, Rivest R L. Introduction to Algorithms. New York: McGraw-Hill, 2009.
- Dale N, Lewis J. Computer Science Illuminated, Sudbury. MA. Jones and Bartlett, 2015.
- Gries D. The Science of Programming. New York, NJ: Springer, 1987.
- Joseph E. Sorting Algorithm: Analysis and Comparison Performance. 2004.
- Kleinberg J, Tardos E. Algorithm Design. Boston, MA: Addison Wesley, 2005.
- Qing X Z, Fang Y, Qian C Z. Introduction to Computer Algorithm Design and Analysis. Beijing, Posts & Telecom Press, 2007.
- Weiss M A. Data Structures and Algorithm Analysis in C++. New Jersey, Addison Wesley, 2014.
- http://en.wikipedia.org/wiki/Algorithm.
- http://msdn.microsoft.com/en-us/library/dd470426.aspx.
- http://whatis.techtarget.com/definition/algorithm.
- http://www.bbc.co.uk/guides/zqrq7ty.
- https://www.khanacademy.org/computing/computer-science/algorithms.

8.8　Summary

- Algorithms are an ordered set of unambiguous steps that produces a result and terminates in a finite time.
- A program is made of sequence constructs, decision constructs, and repetition constructs.
- We usually evaluate algorithms by the following standards: correctness, readability, robustness, time complexity, and space complexity.
- There are many kinds of notations to describe algorithms, such as pseudocode, and UML. UML is a graphical representation of an algorithm. Pseudocode is an informal, English-like representation of an algorithm.
- This chapter introduces the most common algorithms: sorting, and searching.
- A sorting algorithm is an algorithm that puts elements into ordered arrays. Selection sort and bubble sort algorithms are the foundation of the faster sorting algorithms used in computer science today. Merge sort and quick sort are examples of divide and conquer algorithms.

- Searching is a process to find a target in a list of data. There are two basic searching algorithms: sequential search, and binary search. Sequential search can be used in any list, whereas binary search can only be used in the ordered lists.
- There are several algorithms classified by implementation, such as recursion or iteration, serial or parallel or distributed algorithms, deterministic or non-deterministic algorithms, exact or approximate algorithms, and so on.
- There are several algorithms classified by design paradigms, such as exhaustive search, divide and conquer, dynamic programming, and genetic algorithms.

8.9   Practice Set

1. Define the three constructs used in programming.
2. What's the difference between UML and pseudocode?
3. What are the similarities and differences between selection and bubble sort algorithms?
4. What are the two basic searching algorithms? What dissimilarity do they have?
5. What are the performance indices to evaluate an algorithm?
6. The _____ construct would make a judgment.
    a. sequence   b. decision   c. repetition   d. logical
7. The _____ sort need two loops.
    a. selection   b. bubble   c. merge   d. a and b
8. In _____ sort, swap the smallest element from the unsorted list with the element at the beginning of the unsorted list.
    a. selection   b. bubble   c. insertion   d. all of the above
9. In _____ sort, the smallest element moves to the beginning of the unsorted list and there is no one-to-one swapping.
    a. selection   b. bubble   c. merge   d. all of the above
10. _____ is a basic algorithm in which you want to find the location of a target in a list.
    a. Sorting   b. Searching   c. Product   d. Summation
11. _____ is a process in which an algorithm call itself.
    a. Insertion   b. Searching   c. Recursion   d. Iteration
12. Draw a flowchart and write the pseudocode for the bubble sort algorithm.
13. Try to sort the following list using the bubble sort algorithm by hand, and show your work in each pass.

    14   7   23   31   40   56   78   9   2

14. Try to sort the following list using the merge sort algorithm by hand, and show your work in each pass.

    14   7   23   31   40   56   78   9

15. Try to sort the following list using the insertion sort algorithm by hand, and show your work in each pass.

14 7 23 31 40 56 78 9

16. Try to sort the following list using the shell sort algorithm by hand, and show your work in each pass.

48 37 64 95 74 12 26 48 50 3

17. Try to sort the following list using the quick sort algorithm by hand, and show your work in each pass.

14 7 23 31 40 56 78 9

18. A list contains the following elements. The first two elements have been sorted using the selection sort algorithm. Please fill in the list from the third to the last by using the selection sort algorithm.

7 8 26 44 13 23 98 57

19. A list contains the following elements. Using the binary search algorithm to find 88, and show your work in each step. At each step, show the values of first, last, and mid.

8 13 17 26 44 56 88 97

20. A list contains the following elements. Use the binary search algorithm to find 20, and show your work in each step. At each step, show the values of first, last, and mid.

8 13 17 26 44 56 88 97

21. How many comparisons does it take to find 88 and 20 respectively or determine that the item is not in the list by using a binary search algorithm?

8 13 17 26 44 56 88 97

Unit 9 Programming Languages

Language is used by people to describe the real world and to express them, while the computer language is a communication tool between man and computer. On the one hand, people use a variety of computer languages to map the real world they are concerned about to the computer world, but on the other hand, people can create a virtual world that does not exist in the real world by a computer language. In this chapter, we discuss the development, latest ranking, categories, features and common concepts of programming languages.

9.1 Development

Computer language is the medium of communication between man and computer 【计算机语言是人和计算机之间通信的媒介】. A programming language is a formal constructed language designed to transmit instructions【指令】to a machine, particularly a computer. Programming languages have evolved tremendously from machine language【机器语言】, and assembly language【汇编语言】 to high-level languages【高级语言】. Of course, natural language 【自然语言】 is the best programming language and also our final goal. Figure 9.1 presents a time line for computer languages.

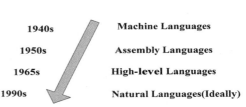

Figure 9.1　Development of computer languages

9.1.1　Machine languages

Machine languages are executed directly by a computer's CPU (Central Processing Unit)【中央处理器(单元)】. They are typically formulated as bit patterns【位模式】, usually represented in octal【八进制】 or hexadecimal【十六进制】. Machine-language programs are written by entering a series of instructions in their binary【二进制】 form which is made of streams of 0s and 1s, and each computer has its own machine language【每类计算机都有自己的机器语言】. The only programming instructions that a computer actually executes are those written using machine language【计算机实际上只能够执行通过机器语言编写的指令】, the instructions built into the hardware【硬件】 of a particular computer. Machine language has the features of flexibility【灵活性】, direct execution【可直接执行】 and

143

high-speed【高运行速度】.

Note: The only language that computer is able to understand is machine language 【计算机能够识别的唯一的语言就是机器语言】.

Table 9.1 shows the machine language program to add two integers.

Table 9.1 Machine language program for adding two integers

Machine language instructions	Finished operation
0001 0000 0010 0000	Get an integer from the memory unit【存储单元】 20, then set it to register【寄存器】A
0011 0000 0010 0001	Add the value of the memory unit 21 to register A【寄存器A中的数值加上内存单元21中的数值,并把和存入寄存器A中】
0010 0000 0010 0010	Store the value of register A to the memory unit 22
0000 0000 0000 0000	Finish the program

9.1.2 Assembly languages

People use English strings, which are easier to read, to represent the machine code【机器码】 and introduce the assembly language. Assembly language uses mnemonic【助记符】letter codes to represent each machine-language instruction. The programmer【程序员】 uses these mnemonics in place of binary digits. Assembly language allows programmers to write instructions in easier ways. However, the CPU cannot understand this symbolic language【符号语言】, so it has to be assembled into machine language before the computer can execute it. Hence, it then developed into the assembly language. The assembly language is a directly invoked processor-oriented【面向处理器的】 programming language.

In assembly language, use of mnemonic instead of machine instruction opcode【操作码】, and address symbol or label instead of instruction or operand address【在汇编语言中,使用助记符而不使用机器指令操作码,使用地址标志或标签而不使用指令或操作数地址】, enhances the readability【可读性】 of the program and reduces the difficulty of preparation.

The process of translating the program in assembly language into machine code is called assembly process【将汇编语言程序翻译成机器代码的过程被称为汇编】, as shown in Figure 9.2.

Figure 9.2 Assembly process

Table 9.2 shows the assembly language program for adding two integers.

Table 9.2 Assembly language program for adding two integers

Assembly language instructions	Finished operation
LOAD X	Get an integer from the memory unit X, then set it to register A
ADD Y	Add the value of the memory unit Y to register A
STORE SUM	Store the value of register A to the memory unit SUM
HALT	Finish the program

9.1.3　High-level languages

Each instruction of the assembly language requires separate coding, which means inefficiency for programming. Thus, in order to improve the efficiency of programmers, and make them focus on applications rather than computer hardware【使程序员只需关注应用程序而不用关注计算机硬件】, the high-level languages were introduced and developed.

The high-level language is not specified for a specific language【高级语言并不是专指一种特定的语言】, but includes many programming languages, such as BASIC (Beginner's All-purpose Symbolic Instruction Code), FORTRAN (Formula Translatior), COBOL (Common Business-Oriented Language), Pascal, C, Visual Basic, Perl, Python, PHP, HTML, FoxPro, Delphi, Visual C, C++, Java, C# (pronounced as C sharp), LISP (List Programming), Prolog (Programming in Logic), and so on. The syntax【语法】 and the command【命令】 format of these languages are not identical.

Table 9.3 shows the Basic language program for adding two integers.

Table 9.3　Basic language program for adding two integers

Basic language statements【语句】	Finished operation
x = 3	Addend x, valuation 3
y = 4	Addend y, valuation 4
Sum = x + y	Store x and y to the sum

9.1.4　Natural languages

The natural language is a language which is used in our everyday life【自然语言就是日常生活中所使用的语言】. It is used for communication, whether by speech, signing or writing. English, Chinese, and Arabic【阿拉伯语】 are examples of natural languages and the sentence "I like all programming languages" is a statement of a natural language. Natural language processing (NLP)【自然语言处理】 is a field of computer science【计算机科学】, artificial intelligence (AI)【人工智能】, and linguistics【语言学】 concerned with the interactions between computers and human (natural) languages.

9.2　Program Translation

The programs written in one of the high-level languages, is called a source program 【用任何一种高级语言编写的程序都叫源程序】. It is saved as text file【文本文件】 called source file【源文件】. The computer doesn't understand and execute such program unless it is translated into the machine language【源程序只有被翻译成机器语言表示的可执行程序后，计算机才能识别和运行】. Translation has two different methods: compilation【编译】 and interpretation【解释】.

Compilation: The compiler【编译器】 is a program that translates the whole source program into a machine language program called an object program【目标程序】. This object program can then be linked with supporting libraries using a linker【连接器】 to create an executable file that runs directly on the CPU. Figure 9.3 shows the translation processes of a compiler.

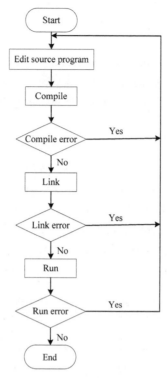

Figure 9.3　Source code compilation process

Interpretation: The interpreter【解释器】 is a program that reads each statement of a source program in sequence, analyzes it what it means, translates it into the corresponding machine code and then immediately executes it, and then goes to the next line. An interpreter is a translating program that translates and executes the statements in sequence【解释器是一段负责顺序地解释和执行语句的程序】.

9.3　Languages' Latest Ranking, Categories and Features

9.3.1　Latest ranking

It is difficult to determine which programming language is most widely used, because every language has its own features. For the same application, C language may need more lines of code (LOC)【代码行】 than Python, but it may be more efficient. Some languages are very popular for specific kinds of applications. For example, COBOL is still widely used in the corporate data center【企业数据中心】, often on large mainframes【主机】; Fortran is used in engineering applications; C is used in embedded【嵌入式】 applications and operating systems【操作系统】; and other languages are regularly used to write specific applications. The TIOBE Programming Community Index has been published since 2001 and updated monthly. The ratings are based on searching the Web with particular phrases that include language names and counting the number of hits【点击数】. Table 9.4 shows the latest ranking in Jun, 2016 and the change in contrast to Jun, 2015.

Table 9.4 Programming languages' ranking

Jun 2016	Jun 2015	Programming Language	Ratings	Change
1	1	Java	20.794%	+2.97%
2	2	C	12.376%	−4.41%
3	3	C++	6.199%	−1.56%
4	6	Python	3.900%	−0.10%
5	4	C#	3.786%	−1.27%
6	8	PHP	3.227%	+0.36%
7	9	JavaScript	2.583%	+0.29%
8	12	Perl	2.395%	+0.64%
9	7	Visual Basic .NET	2.353%	−0.82%
10	16	Ruby	2.336%	+0.98%
11	11	Visual Basic	2.254%	+0.41%
12	23	Assembly Language	2.119%	+1.36%
13	10	Delphi/Object Pascal	1.939%	+0.07%
14	14	Swift	1.831%	+0.39%
15	5	Objective-C	1.704%	−2.64%
16	13	R	1.540%	+0.02%
17	15	MATLAB	1.447%	+0.01%
18	17	PL/SQL	1.346%	+0.12%
19	26	D	1.063%	+0.45%
20	18	COBOL	1.048%	+0.10%

In addition to the 20 most common languages in the table above, there are also many famous languages, for example, Fortran, Scratch, ABAP (Advanced Business Application Programming)【高级企业应用编程语言】, Dart, SAS, Groovy, T-SQL (Transact-SQL), Lisp, Ada, Scala, Lua, Logo, Prolog, Scheme, RPG, LabVIEW, OpenEdge ABL (OpenEdge Advanced Business Language), Erlang, Haskell, Alice, Apex, Bash, F#, Q, Ladder Logic, Rust, Awk, Go, FoxPro, VBScript, TypeScript, and so on. We will not go into details on other languages.

9.3.2 Programming language paradigms

Computer languages can be divided into four categories according to the approach they use to solve a problem: procedural (imperative) languages【过程式（命令式）语言】, object-oriented languages【面向对象语言】, functional languages【函数式语言】, and declarative languages【声明性语言】. Table 9.5 shows the categories of programming languages with examples.

Table 9.5 Categories of programming languages

	Procedural	FORTRAN/COBOL/BASIC/C/Pascal/Ada
Computer language paradigms	Object-oriented	Smalltalk/C++/Visual Basic/C#/Java/Python/Ruby/JavaScript/Perl/PHP/Simula/Lua/R/Objective C/TCL/Io
	Functional	LISP/Scheme
	Declarative	Prolog/Erlang

Procedural (imperative) languages

Procedural programming【过程式程序设计】, derived from structured programming【结构化程序设计】, is based upon the concept of procedure-call【过程调用】. A procedural program is composed of one or more units or procedures or modules【一个过程式程序是由一个或多个单元、程序或模块组成的】. Procedures are also called functions【函数】, routines【程序】, subroutines【子程序】, or methods【方法】. They contain a series of computational steps to be executed. During a program's execution, any given procedure might be triggered (called) at any point, by other procedures or itself.

Typical examples of procedural languages are FORTRAN, COBOL, BASIC, C, Pascal and Ada.

Object-oriented languages

The object-oriented programming (OOP)【面向对象程序设计】 handles some interacting objects【交互对象】. Each object is responsible for its own actions【每个对象只对自己的行为负责】. In the procedural programming, data items are considered passive and are manipulated by the program【数据项被认为是被动的，并且是由程序操纵的】. While in object-oriented programming, data objects【数据对象】 are active. They have their own actions and data, and the execution depends on them. Objects and the action to be performed on these objects are bundled together, thus making each object responsible to perform one of the actions when it receives the appropriate stimulus by some external events【事件】. These languages allow the programmers to implement algorithms【算法】 as a hierarchy of objects.

Object-oriented programming achieves three main objectives of software engineering【软件工程】: reusability【可重用性】, flexibility, and scalability【可扩展性】. In order to standardize the behaviors of objects【为了规范对象的行为】, each object can receive information, process data and send information to other objects. The design concepts of object-oriented programming include objects, classes【类】, methods, data abstraction【数据抽象】, inheritance【继承】, dynamic binding【动态绑定】, data encapsulation【数据封装】, polymorphism【多态性】, and message passing【消息传递】. These concepts concretely express the idea of object-oriented programming.

Typical examples of object-oriented languages are Smalltalk, C++, Visual Basic, C#, Java, Python, Ruby, JavaScript, Perl, PHP, SIMULA, LUA, Objective C, TCL, Io and R.

Functional languages

Functional languages define programs and subroutines as mathematical functions【函数式语言将程序和子程序都定义为数学函数】. In this context, a function is a black box【黑盒】 that maps inputs into outputs, depending on the arguments【参数】 that are input to the function. Computation is expressed as functions. The solution to a problem is expressed as function calls【函数调用】. Figure 9.4 shows a function in a functional language.

Typical examples of functional languages are LISP and Scheme.

Figure 9.4　A function in a functional language

Declarative languages

Declarative languages describe the logic of a computation rather than expressing its control flow 【声明式语言仅仅描述计算的逻辑而不是表达控制流】. This is in contrast with procedural (imperative) programming, which needs an explicitly provided algorithm and has to pass consecutive orders 【连续指令】 to a computer. In fact, all (pure 【纯的】) functional and logic-based programming languages are declarative.

Typical examples of declarative languages are Prolog and Erlang.

9.3.3 Features of programming languages

Table 9.6 shows the features, application scenarios and development tools of some popular programming languages.

Table 9.6 Programming languages

Language	Features	Application Scenarios	Development Tools
Java	Object-oriented; Cross-platform【跨平台】； Multi-thread【多线程】； Powerful class library【类库】	All kinds of web development, especially large enterprise applications	Eclipse/NetBeans/JBuilder
C	Procedure-oriented; High performance【高性能】； Strong expression ability; High portability【可移植性】； Dynamic memory management【动态内存管理】	Most of the operating systems, drivers 【驱动程序】 and the underlying software	VC6.0/Visual Studio/CodeBlocks/Gcc
C++	Object-Oriented; Flexible【灵活性】； Compatible【兼容性】； High performance; Standard template library【标准模板库】; Dynamic memory management	Large client applications, game development, and the underlying framework	VC6.0/Visual Studio/Code:Blocks/G++
C#	Object-oriented; High scalability; Garbage collection【垃圾回收器】；	Closely integrated with the Web	Visual Studio/Mono Develop
Python	Simple; Clear; Powerful class library【类库】	Daily need of small tools, including script task of system administrators	PyCharm
PHP	Sever-Based; Perfect combination with Apache and MySQL	Server development【服务器开发】, such as Unix, Linux, Windows	Appserv/phpStudy
Visual Basic.NET	Easy to learn and use; Visual-design【可视化设计】； Powerful class library; Interactivity【交互性】	Multimedia【多媒体】	Visual Studio
JavaScript	Simple; Secure; Dynamic; Cross-platform	Almost all the fields, such as Web and mobile development【移动开发】	Spket/Ixedit/KomodoEdit/EpicEditor

续表

Language	Feature	Application Scenarios	Development Tools
Perl	Complex and flexible data structure; Simple; Powerful pattern matching【模式匹配】	Many Standard and third-party modules, Graphical user interfaces (GUIs)【图形用户界面】development	WxPerl/ActivePerl/PerlBuilder/Padre
Ruby	Cross-platform; Flexible; Scalable; Object-oriented; Powerful rails	Rapid web development	RDT (Ruby Development Tools)/RadRails

Some important programming languages for specific areas are given below:

1. Mathematical Analysis and Processing: MATLAB, R, SAS

MATLAB (matrix laboratory)【矩阵实验室】 is a business mathematical software produced by the American MathWorks company. It can achieve algorithm development, data visualization【数据可视化】, data analysis【数据分析】 and numerical calculation【数值计算】 of senior technical computing language and interactive environment【交互式环境】, mainly including two most of MATLAB and Simulink.

R is a complete set of data processing【数据处理】, calculation and drawing software system. It has some unique features that make it very powerful and simple, such as data processing, array【数组】 operation, statistical analysis【统计分析】, and statistical graphics【统计图形学】.

SAS (Statistics Analysis System)【统计分析系统】 is mainly used for statistical analysis. It can read in data from common spreadsheets【电子表格】 and databases and output the results of statistical analyses in tables, graphs, and as RTF (Rich Text Format)【富文本格式】, HTML, and PDF (Portable Document Format)【便携式文档格式】 documents.

2. Parallel Computing Model【并行计算模型】: CUDA, OpenACC, OpenMP, OpenCL

There is no special language for parallel programming【并行编程】, but a lot of languages are used in parallel computing model, such as CUDA (Compute the Unified Device Architecture)【统一计算设备架构】, OpenACC (Open Accelerators), OpenMP (Open Muti-processing)【共享存储并行编程】, OpenCL(Open Computing Language)【开放计算语言】, and so on.

3．Database: SQL (Structured Query Language)【结构化查询语言】, PL (SQL), Transact-SQL

SQL is a standardized computer language that was originally developed by IBM for querying, altering and defining relational databases【关系数据库】, using declarative statements【声明语句】.

T-SQL (Transact-SQL) is Microsoft's and Sybase's proprietary extension to SQL and is central to using Microsoft SQL Server.

PL/SQL (Procedural Language/Structured Query Language) is Oracle's procedural extension to SQL and the Oracle relational database.

4. Designing for the OS X and iOS operating systems: Object-C, Swift

Native application【原生应用程序】 of iOS must be written in Swift or Objective-C, whereas some elements can use C or C++ optionally.

5. Artificial Intelligence: Prolog, LISP

Prolog (Programming in Logic)【逻辑程序设计语言】 is a general purpose logic programming language most widely used in artificial intelligence. Its programming method is more like using logic to describe programming, so it can develop more quickly than any other language.

LISP is a generic high-level computer programming language. It is one of the oldest high level programming languages and continues to be popular in the field of artificial intelligence to the present day.

6. Automatic Testing【自动化测试】: TCL (Tool Command Language)

TCL is mainly used for individual and embedded companies such as Huawei and Maipu.

7. Web Page Development【网页开发】: HTML, HTML5

HTML (Hypertext Markup Language)【超文本标记语言】 is the standard markup language used to create web pages, along with CSS (Cascading Style Sheets)【层叠样式表】, and JavaScript.

HTML5 is a markup language used for structuring and presenting content on the World Wide Web【万维网】. It is the fifth and current version of the HTML standard. HTML5 can undertake independent game development.

9.4 Common Concepts of Programming Languages

In this section, we introduce some basic elements and common concepts in a programming language. All the examples in this section are based on C language.

Identifiers【标识符】: An identifier is used to name an object. Identifiers allow us to reserve memory locations that can be referenced by the identifier's name in our program.

Data Type【数据类型】: A data type defines a set of values and the set of operations on those values【数据类型定义了一组值和关于这些值的操作的集合】. For example, in C language, there are five basic data types: int【整型】, char【字符型】, float【(单精度)浮点型】, double【双精度浮点型】, and void. Apart from the simple data types, C language also defines arrays, pointers【指针】, structures【结构体】, unions【共用体】, and functions.

Variables【变量】: Variable is a named memory location associated with an identifier, which contains some known or unknown stored value, as shown in Figure 9.5. The value of variables can change during runtime.

Variable declarations【声明】: A declaration is a statement that associates a type and name (identifier) with a variable, an action, or some other entity within the language that can be given a name so that the programmer can access that item, object, or process by name. Most programming languages require that variables must be declared prior to use【大多数语言中变量都需要在使用前先声明】. The computer allocates the required storage space and names it.

Constants【常量】: A constant is a named memory location with an associated value, but the value cannot be altered during the execution after it has been defined at the beginning of the program. For example in C or C++ program, π can be defined at the beginning and used during execution of the program, such as const float pi = 3.14.

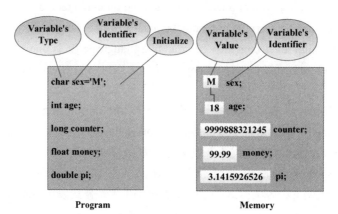

Figure 9.5   Variables

The type of constant must be defined when it is declared.

Input and Output【输入输出】: Input is used to read data from other devices (e.g. monitor and file). Output is used to write data to other devices. Every language usually provides some input/output functions or interfaces【接口】.

Expressions【表达式】: An expression defines a set of operands【操作数】 and operators【运算符】. These operators include arithmetic operators【算术运算符】, relational operators【关系运算符】 and logical operators【逻辑运算符】. Table 9.7, Table 9.8, and Table 9.9 show some operators used in C, C++, and Java.

Table 9.7   Arithmetic operators

Operator	Definition	Example
+	Addition【加法运算符】	7 + 11
−	Subtraction【减法运算符】	8−3
*	Multiplication【乘法运算符】	a * 2
/	Division【除法运算符】 (the result is the quotient)	a / b
%	Division【取余运算符】 (the result is the remainder)	20 % 3
++	Increment【递增运算符】 (add 1 to the value of the variable)	i++
——	Decrement【递减运算符】 (subtract 1 from the value of the variable)	i——

Table 9.8   Relational operators

Operator	Definition	Example
<	Less than	count < 6
<=	Less than or equal to	count<= 6
>	Greater than	count > 5
>=	Greater than or equal to	count >= 5
==	Equal to	a == (b + c)
!=	Not equal to	a != b

Table 9.9 Logical operators

Operator	Definition	Example
!	Not【逻辑非】	! (a > b)
&&	And【逻辑与】	(a < 3) && (b > 6)
||	Or【逻辑或】	(a < 7) || (b > 15)

Statements: A statement is a command that causes a specific action to be performed by the program【一条语句就是能够引发程序执行特定行为的一条指令】. In C language, an expression becomes a statement when it is followed by a semicolon. Statements are executed in sequence and do not have values【语句是按序执行的并且没有返回值】. They are translated directly into one or more executable computer instructions. Figure 9.6 shows the categories of the statements.

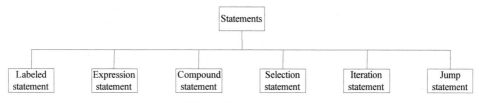

Figure 9.6 Statements

Functions: Function is an independent module【模块】 that is called to do specific tasks. In some programming languages, a function is equivalent to a subroutine, or a procedure. For example, a function is similar to a subroutine or function in FORTRAN, or a method or member function【成员函数】 in most object-oriented languages, or a procedure or function in C and Pascal. A function provides a convenient way to encapsulate some computation, which can then be used without caring about its implementation. Sometimes, knowing the execution result is enough. Therefore, with properly designed functions, it is possible to ignore the execution process of a job. Figure 9.7 is an example of function declaration.

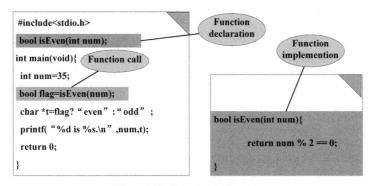

Figure 9.7 Function declaration

9.5 Severe Software Errors

In the programming process, you must be thoughtful to avoid various mistakes. Otherwise, those mistakes may cause disastrous consequences【灾难性的后果】. Here are some examples of notorious bugs【臭名昭著的漏洞】 which have bad influence at that time.

Bug 1: Mariner I space probe【水手一号航空探测器】 **(1962)**

While transcribing a handwritten formula【公式】 into a computer navigation code【导航代码】, a single bar was left out.

Consequence【结果】: The vessel【宇宙飞船】 went so far off the mission control【太空航行地面指挥中心】 that it had to be destroyed. An $18.5 million space exploration vehicle【太空探索飞船】 was lost.

Bug 2: AT&T【美国电话电报公司】 **network outage**【网络中断】 **(1990)**

A bug in a new release of the software that controls AT&T's #4ESS long distance switches【远程交换机】 causes these mammoth computers【大型计算机】 to crash【死机】.

Consequence: Leaving an estimated 60 thousand people without long distance service【远程服务】 for nine hours. AT&T lost some $60 million to $75 million.

Bug 3: Pentium chips【奔腾芯片】 **math coprocessor**【数学协同处理器】 **error (1993)**

The built-in divider【内置除法器】 in the Pentium FPU (floating point unit)【浮点运算单元】 had a flawed division table【有纰漏的除法表】.

Consequence: Intel's Pentium chip occasionally made mistakes beyond the eighth decimal point【在小数点8位之后出错】.

Bug 4: Explosion of the Ariane 5 rocket 【亚利安五号运载火箭】 **(1996)**

The conversion【转换】 of a 64-bit floating point to a 16 bit signed integer【有符号整数】 was failed because the floating point number【浮点数】 was larger than 32,767, which is the largest integer storable【可以存储的】 in a 16 bit signed integer.

Consequence: The rocket exploded just 40 seconds after lift-off on its maiden voyage【首航】.

Bug 5: Mars climate orbiter【气象卫星】 **crashes (1998)**

The use of imperial units【英制单位】 instead of metric system【公制】 in the orbiter by the sub contractor【分包商】 causes this accident.

Consequence: The crash of the $125 million spacecraft on Red planet【火星】.

Bug 6: Millennium Bug or Y2K【千年虫】 **(1999)**

Two digits【两位数】 were used to store the year for dates, so year 2000 could only be represented as '00', which might confuse computers into misinterpreting【误解/曲解】 '00' as year 1900 rather than 2000.

Consequence: Many devices containing computer chips【计算机芯片】 were believed to be at risk so preparations for the Y2K bug had a significant cost and time impact on all industries.

Bug 7: Software-related radiation therapy【放射治疗】 **failure (2001)**

A bug in the Multidata【多数据】 software caused the patients to be over-radiated and the process went on for seven months.

Consequence: Cobalt-60 machine in Panama's National Cancer Institute【巴拿马国家癌症研究所】 overdosed more than two-dozen patients with gamma radiation【伽马辐射】, which claimed several lives.

Bug 8: Disastrous power outage【停电事故】 **in Northeastern America and Southeastern**

Canada (2003)

The operator turned off alarm caused by a bug provoked by the power flow monitoring tool【监测工具】.

Consequence: Power lines were dead in America and the international connections between Canada and the United States went offline【脱机】 resulting in 55 million people without power.

Bug 9: Michigan Department of Corrections grants prisoners early release (2005)

The register incorrectly reported early release of 23 prisoners due to a computer programming glitch【程序故障】.

Consequence: An undisclosed number of inmates were also kept in jail past their release dates.

Bug 10: Shutdown【停机】 of the Hartsfield-Jackson Atlanta International Airport (2006)

A software malfunction【软件故障】 failed to alert the security screeners【安检员】 that the image of a suspicious device【可疑的设备】 was just part of a routine test【例行测试】.

Outcome: The airport authorities evacuated the security area for two hours while searching for the suspicious device, causing more than 120 flight delays, and forcing many travelers to wait outside the airport.

Bug 11: L.A. Airport flights grounded 【航班停飞】(2007)

Due to a single piece of faulty embedded software【嵌入式软件】 on a network card【网卡】, incorrect data was sent out on the U.S. Customs and Border protection network【保护网】.

Consequence: More than 17,000 planes were grounded for the duration of the outage.

Bug 12: Google accidently blocks the Internet (2009)

One of Google's programmers was adding websites to the malware registry【恶意注入】 when he accidentally entered / instead of a full URL.

Consequence: For close to an hour, every website was flagged as possibly harmful, and Google blocked all users from visiting those suspicious sites, which were all the sites everywhere, including Google's own pages.

Bug 13: Tax returns【纳税申报单】 hit by HMRC(Her Majesty's Revenue And Customs)【英国税务及海关总署】 software bug (2010)

HM Revenue & Customs (HMRC) did not fix a software bug for several months.

Consequence: About 6.8 million individuals either underpaid tax【偷税】 or got refunds due to tax code problems.

Bug 14: Heartbleed Bug【致命错误】 in OpenSSL（Open Secure Sockets layer）【开放式安全套接层协议】 (2011)

Someone inserted the Heartbleed bug into OpenSSL which leaked a small slice of memory【泄露了一小块内存】 from Web servers【Web 服务器】 and client PCs.

Consequence: For two years this bug nabbed passwords, card details, and even encryption keys.

Bug 15: Apple's Maps app for iOS 6 geographic errors【地理错误】 (2012)

Building a map is extremely difficult and requires not only smart software but also thousands of man-hours of handwork. iOS 6 maps was a patchwork and lacked precision【精度】.

Consequence: The Apple's iOS 6 Maps app misidentified geographic locations【地理位置】, and several major landmarks【地标】 were labeled inaccurately or wildly misplaced.

Bug 16: BOE Software Defect (2013)

The development team【开发团队】 released software【软件发布】 with a known defect【缺陷】 in Bank of England, which prevented trading in two of the most closely watched indexes【观测指标】 in the U.S.

Consequence: Ordered to pay $6 million in fines【处罚】 and made changes to system and software.

Bug 17: Year 2038

UNIX operating system handles dates by counting how many seconds a date is since 1st of January 1970 due to size limit of signed 32-bit integer.

Consequence: UNIX-based embedded systems【基于 UNIX 的嵌入式系统】 can only handle dates up to 19th of January 2038.

These examples tell us: Programmers or designers must do their jobs with great care.

9.6 References and Recommended Readings

For more details about the subjects discussed in this chapter, the following books are recommended:

- Cooke D A. Concise Introduction to Computer Languages. Pacific Grove, CA: Brooks/Cole, 2003.
- Tucker A, Noonan R. Principles and Paradigms. Burr Ridge, IL: McGraw-Hill, 2002.
- Pratt T, Zelkowitz M. Programming Languages, Design and Implementation. Englewood Cliffs, NJ: Prentice Hall, 2001.
- Sebesta R. Conceptes of Programming Languages, Boston, MA: Addison Wesley, 2006.
- Kernighan B W, Ritchie D M. The C Programming Language. Upper Saddle River, NJ: Prentice Hall, 1988.
- Deitel P, Deitel H. C How to Program, Pearson Education Limited, 2016.
- Tucker A B, Noonan R E. Programming Languages: Principles and Paradigms. McGraw-Hill, 2006.
- Friedman D P, Wand M. Essentials of Programming Languages. Cambridge, MA: MIT Press, 2008.
- Orchard D, Rice A. Computational Science Agenda for Programming Language Research. Procedia Computer Science, 29: 713-727, 2014.
- www.tiobe.com.

9.7 Summary

- The only language a computer can understand is machine language.
- An assembly language uses symbols to represent various machine language instructions.

Assembly languages are also called symbolic languages.
- A high-level language focuses on problem-solving. C, C++, Pascal, Python, LISP, Prolog, FoxPro, Delphi and VC are all high-level languages.
- Two methods are used for source program to object program's translation: compilation and interpretation.
- There are four categories of computer languages: procedural (imperative), object-oriented, functional, and declarative.
- A procedural program is composed of one or more units or modules. FORTRAN, COBOL, BASIC, C, Pascal, and Ada are all procedural languages.
- In object-oriented programming, the objects and the operations to be applied to them are tied together. Smalltalk, C++, Visual Basic, C#, and Java are all object-oriented languages.
- Functional programming languages define programs and subroutines as mathematical functions. LISP and Scheme are functional languages.
- Declarative languages describe a problem rather than defining a solution. Prolog is a declarative language.
- Native application of iOS must be written in Swift or Objective-C.
- Some common concepts in procedural language include identifier, data types, variable, constant, input and output, expression, statement, and function.
- In the programming process, you must be thoughtful to avoid various mistakes.

9.8　Practice Set

1. Differentiate between machine language and assembly language.
2. What are the differences between compilation and interpretation?
3. What are the four categories of computer languages? Give an example of each.
4. How is procedural programming different from object-oriented programming?
5. What is a functional language? Give an example of a function language.
6. What is a variable?
7. What are arithmetic operators, relational operators and logical operators in the procedural languages?
8. Do some research and learn the basic essentials of object-oriented programming languages. You can take the example of C++ and Java.
9. If you like programming, you can try to use a language to solve some real problems. For example, you can use Python to calculate n.
10. Do you think what functions should the programming language supply in the future?
11. List steps in programming language compilation.
12. The only language which a computer can directly run is _____ language.
    a. symbolic　　　b. assembly　　　c. high-level　　　d. machine
13. FORTRAN, COBOL and BASIC can be classified as _____ languages.

  a. procedural b. functional c. declarative d. object-oriented

14. _____ is a program's code in high-level language.

  a. An object program   b. A source program

  c. A machine code   d. None of these

15. _____ looks at each line of your program, works out what it means, obeys it, and then goes onto the next line.

  a. Assembler b. Compiler c. Translator d. Interpreter

16. A (n) _____ program is composed of one or more units or modules.

  a. procedural b. functional c. declarative d. object-oriented

17. Pascal is a(n) _____ language.

  a. procedural b. functional c. declarative d. object-oriented

18. C++ is a(n) _____ language.

  a. procedural b. functional c. declarative d. object-oriented

19. LISP and Scheme are both _____ languages.

  a. procedural b. functional c. declarative d. object-oriented

20. An example of a(n) _____ language is Prolog.

  a. procedural b. functional c. declarative d. object-oriented

Unit 10 Software Engineering

As a way of hedging against software crisis【软件危机】, we need to put emphasis on software engineering【软件工程】 and lifecycle of the software. We will show models used for the development process in this chapter followed by a brief discussion of the phases in the development process. In the end, we will introduce software process management and CMMI (Capability Maturity Model Integration)【软件能力成熟度集成模型】, and the importance of quality, modularity【模块化】 and documentation【文档】.

10.1 Introduction

10.1.1 Software crisis

Software crisis refers to the difficulty of writing useful, efficient, correct, and verifiable computer programs within the required time and budget【预算】. With the rapid increase in the complexity of the large-scale software【大型软件】 and the hardware development process, "software crisis" emerged in several ways because of issues such as over-budget, over-time, software inefficiency, software's low quality, etc. The basic problems during software development process are briefly summed up as follows.

（1）To develop flawless【没有缺陷的】 software is a big challenge.

（2）A significant amount of work and time is needed in removing flaws or bugs【漏洞】 from the end product【最终产品】.

（3）The functionality of the software does not always perfectly match with the requirements of the end-users【最终用户】.

（4）It is very hard to maintain software, despite being provided codes with lots of comments【尽管可以在代码中添加大量注释，维护软件仍然是一件困难的事】. It is quite possible that the software maintainers differ from the core developer【核心开发者】. So the ability to understand codes decrease rapidly over time.

Therefore, in the late 1960s, the necessity of software engineering was also emerged. Software engineering is the application of principles to obtain reliable software and solve software crisis.

10.1.2 The Goals of software engineering

The final goal of software engineering is to create high-quality software【高质量软件】that meets requirements of end-users within the decided schedule and with least cost.

Possible goals of software engineering are as follows.

（1）Create software that uses fewer resources【资源】 (cheaper and faster【便宜且快捷】).

（2）Create software that is easily modifiable for unanticipated requirements【无法预知的需求】.

（3）Create software that meets customer's requirements.

（4）Create software that is of high quality.

（5）Create software that is easily understandable for developers.

（6）Create software that is easily usable for end-users.

（7）Create software that is free of bugs and errors.

（8）Create software that is efficient, flexible, extensible, re-usable, reliable, correct, and readable.

10.1.3 Lifecycle of software

The software lifecycle【软件生命周期】 is a fundamental concept in software engineering. It means a series of changes that software goes through from the beginning to its outdated phase (Figure 10.1).

No matter how well software has been developed and implemented, it still may contain bugs. Modifications are often required to correct errors and fix problems found in the software.

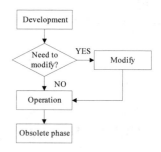

Figure 10.1　The software lifecycle

10.2　Software Development Life Cycle

The software development life cycle (SDLC) involves four phases: analysis, design, implementation, and testing.

10.2.1 Analysis phase

The analysis phase【分析阶段】 is the first one of development process. The result in this phase is the requirement specification document【需求说明文档】 that shows the function and scope of the target software. This phase can use two different approaches: procedure-oriented analysis and object-oriented analysis. The selection of approach depends on the category of the programming language planned to be used in the implementation phase.

1. Procedure-oriented analysis (Structured or Classical analysis)

Procedure-oriented analysis【面向过程的/结构化的/经典的分析】is used if a procedural language will be used in the implementation phase【实现阶段】. Several modeling tools are used for specification, such as Data Flow Diagrams (DFD)【数据流图】, Entity-Relationship Diagrams (ERD)【实体关系图】,

and state diagrams【状态图】.

DFDs show the data movement in the system. Figure 10.2 shows a DFD for an online reservation system of a small hotel.

The process, handle reservation【预定】, accepts or denies a reservation request by the potential guest based on availability of rooms.

ERDs【实体关系图】 are a graphical representation of entities【实体】 and their relationships【关系】. They are also used in database design. ERDs use specialized symbols to represent three different types of information. Boxes represent entities. Diamonds represent relationships and ovals represent attributes【属性】. Figure 10.3 shows an entity-relationship diagram.

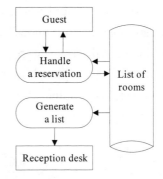

Figure 10.2 An example of a data flow diagram

State diagrams provide information of the state of the entities depending on events in the system. Figure 10.4 shows a state diagram for the working of a one-passenger elevator. When the passenger press a floor button, the elevator moves to the requested floor. It do not process any request until it reaches its destination.

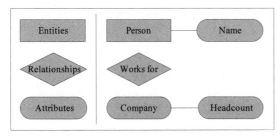

Figure 10.3 An entity-relationship diagram

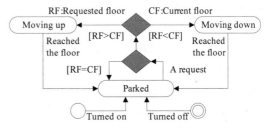

Figure 10.4 An example of a state diagram

2. Object-oriented analysis

Object-oriented analysis【面向对象分析】is used if the implementation phase uses an object-oriented language. Some of the tools used for the specification document in object-oriented analysis are as follows.

Use case diagram

Use case diagram【用例图】 is one of behavioral UML (Unified Modeling Language)【统一建模语言图】 diagrams. A UML diagram provides the way to understand and describe requirements of a system. It has four components, system【系统】, use cases【用例】, actors【角色】 and relationships【关系】. A system, denoted by a rectangle, performs a function. Use cases, denoted by rounded rectangles, are lists of actions or events, which define the interactions between roles and systems. Use case analysis is one of valuable and important techniques in modern software engineering. Figure 10.5 shows the UML diagram for the working of a one-passenger elevator.

Class diagram

Class diagram【类图】 is one of structural UML diagrams. A class diagram provides the way to understand and describe the structure of a system. For example, we have two entity classes【实体类】in simple elevator implementation: the buttons and the elevator itself. Therefore, we have a button class and an elevator class. You may notice that buttons have two types: the elevator buttons in the hallways and the floor buttons inside the elevator. The button class has two subclasses that inherit【继承】from the button class as the parent class【父类】: the elevator button class and the floor button class. Figure 10.6 shows the class diagram for the working of a one-passenger elevator.

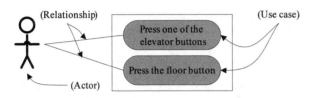

Figure 10.5　An example of a use case diagram

Figure 10.6　An example of a class diagram

10.2.2　Design phase

Design phase【设计阶段】 defines how the system will accomplish results defined in the analysis phase. All components of the system are described in this phase, including desired features, and operations. This phase can use two different approaches: procedure-oriented design【面向过程设计】and object-oriented design【面向对象设计】.

1. Procedure-oriented design

In procedure-oriented design, both procedures and data are designed. The whole system is divided into a set of modules or procedures. Some of the tools and techniques used for the modules in procedure-oriented design are as follows.

Structure charts

Structure charts【结构图】 are a useful tool to illustrate the relations between modules【模块】 in procedure-oriented design. Figure 10.7 shows the structure chart for the working of a one-passenger elevator.

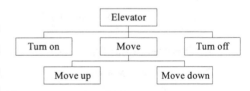

Figure 10.7　An example of a structure chart

Modularity

Modularity means dividing a large task into smaller and understandable tasks without any context change. Figure 10.7 shows the modularity in the elevator system. There are two main concerns about modularity in a system: coupling 【耦合度】and cohesion【内聚度】.

Coupling and cohesion

Coupling is a measure of tightness between modules. Coupling between modules is inversely proportional to their independency. Loosely coupled modules are reusable, less error-prone【不易于出错的】, and easily modifiable. Therefore, a software system having minimum coupling is desirable.

Cohesion is a measure of relativeness【相关性】 between modules. A software system having maximum possible cohesion is desirable. To maximize the modularity of the software, you need to minimize independence and maximize cohesion between modules in a software.

2. Object-oriented design

The concept of class is one of core ideas in object-oriented design. Classes are made of attributes and methods【方法】. The details of these attributes and methods are described in this phase. Figure 10.8 shows the classes for the working of a one-passenger elevator.

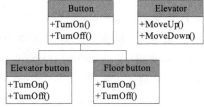

Figure 10.8 An example of classes with methods

10.2.3 Implementation phase

In the implementation phase, the project team【项目团队】chooses a language, and then actual coding of the software system starts. C++ and Java are commonly considered languages that implement object-oriented design. There is a far more important issue than the selection of a language to implement a software. It is software quality【软件质量】A high quality software system must be accurate, efficient, reliable, secure, usable, changeable, quickly recoverable, flexible, testable【可测试的】, re-usable【可重用的】, interoperable【可互操作的】, and portable. Such a system meets the users and organization's requirements and standards, and runs efficiently on the platform for which it was developed. Some of the attributes of quality are operability【可操作性】, maintainability【可维护性】 and transferability【可移植性】.

10.2.4 Testing phase

The objective of testing phase【测试阶段】 is to find and eliminate potential errors【潜在的错误】in the software. Software testing is the process of ensuring that it is actually fulfilling user's requirements gathered during requirement analysis【需求分析】. The objective of this phase is to assess the functionality and correctness of the software. Software debugging【软件调试】 should be go side-by-side with software testing. Software debugging is the process of finding and removing errors by re-designing or re-coding. Many types of software testing methods【软件测试方法】 are employed, depending on the type of the software being developed.

Unit testing 【单元测试】 is a commonly used method. It tests the basic pieces such as class and methods. It tests program parts in isolation from the rest of the system【单元测试能将程序要测试的部分与其他部分隔离开来】. There are two kinds of unit testing: white-box testing and black-box testing (Figure 10.9). White-box testing is a program-based testing【基于程序的测试】, and black-box testing is specification-based testing【基于说明文档的测试】.

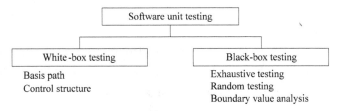

Figure 10.9 Software unit testing

1. White-box testing

White-box testing【白盒测试】(or glass-box testing【玻璃盒测试】) is one of audit trails【审计跟踪】through a computer. To do this, white-box testing assumes that the tester【测试人员】knows everything about the software. In other words, the software is like a glass box where everything inside the box is visible. A dedicated team of software engineers performs this testing. It uses the structure of the software. Therefore, it needs to meet the following four requirements:

（1）All independent paths in each module need to be tested at least once.

（2）All the decision constructs need to be tested on each branch.

（3）Each loop construct needs to be tested.

（4）All data structures need to be tested.

There are many white-box testing techniques. We briefly discuss two of them: basis path testing【基础路径测试】and control structure testing【控制结构测试】.

- Basis path testing

Basis path testing is one of white-box testing techniques proposed by Tom McCabe. In this technique, each statement【语句】in the software is executed at least once.

- Control structure testing

Control structure testing includes three testing methods: condition testing, data flow testing, and loop testing.

Condition testing

Condition testing【条件试验】applies to every condition expression【条件表达式】in the module to check its correctness. A simple condition is a relational expression【关系表达式】, while a compound condition【复合条件】is a combination of simple conditions and logical operators【逻辑算子】.

Data flow testing

Data flow testing【数据流测试】tests the flow of data in each module. It also checks the value of variables in all the assignment statements【赋值语句】.

Loop testing

Loop testing【回路测试】tests the validity【有效性】of all types of loops.

2. Black-box testing

Black-box testing【黑盒测试】is to audit around the computer. The software is like a black box into which the tester cannot see. You only need to focus on the results in black-box testing, other than the processes. In other words, it is not important what is inside it and how it works. Black-box testing only tests software's inputs and outputs. There are several techniques used for black-box testing: exhaustive testing 【穷举测试】, random testing【随机测试】, and boundary-value testing【边界值测试】.

- Exhaustive testing

Exhaustive testing is a testing method to test the software for all possible values in the input domain【输入域】. Generally, software is so complex, and the input domain is so huge, that it is often impractical to do so. So exhaustive testing may be used in small scale.

- Random testing

In random testing【随机测试】, programs are tested by random selection of a subset of input values from the input domain. It is critical how the subset is chosen among the domain input values. For random selection of input values, random number generators are useful.

- Boundary-value testing

Boundary-value testing is the testing of boundary values【边界值】 that usually cause errors. If an input value is in the range of 50-100, then 50 and 100 are the boundary values. For example, a condition in a module may be coded as num < 50, instead of num ≤ 50.

10.3　Software Development Models

Software development models are the various methodologies【方法论】 or processes that are developed in order to achieve the project's objectives. There are several models for the development process. Some of the most important and common models are waterfall model【瀑布模型】, incremental model【增量模型】, prototyping model【原型模型】, spiral model【螺旋模型】, agile model【敏捷软件开发模型】, iterative model【迭代模型】, V model, and rapid application development (RAD) model【快速程序开发模型】. A brief description of some of these models is given here.

10.3.1　Waterfall model

In the waterfall model, the development process has a linear sequential flow. This means that a phase is started only after the previous phase is completed. In this model, progress is seen as flowing steadily downwards (like a waterfall) through several phases (Figure 10.10).

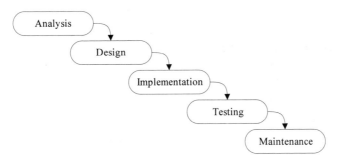

Figure 10.10　The waterfall model

This model is used only when requirements are clearly defined and understood. Waterfall model is simple, easy to understand, easy to use, easy to manage, easy to explain to users, have well defined activities and stages, and each phase has specific deliverables【可交付物】. The team who works on the design phase has the complete results of the analysis phase, so they know exactly what to do. Similarly, in testing phase, team members have the complete results of their previous phase, so they can test the whole system. However, in this model, it is very difficult to go back to a previous stage and change something. It is uncertain and poor model for long projects where requirements are at a high risk of changing. Another disadvantage of waterfall model is difficulty in locating a problem. If there is a problem in part of the process, the entire process must be checked.

10.3.2　Incremental model

In the incremental model, software is developed by breaking a project into a series of steps, thus providing more ease-of-change during the development process. It is a sequential model in which a simple version of the entire system is first developed. More details are added in the second version for another testing, while some are left unfinished. When the existing version works properly, new version is developed with some added functionality. This series of processes continues until all required functionalities have been added. The concept of incremental model is shown in Figure 10.11. Incremental model is easier to manage, more flexible and less costly to modify requirements and scope. Testing and debugging is easier in smaller versions or iterations【迭代】. However, partitioning the entire system and then integrating between iterations can be problematic.

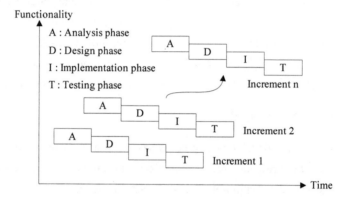

Figure 10.11　The incremental model

10.3.3　Prototyping model

The main idea of prototyping model is the creation of prototypes, i.e., incomplete versions of the software program being developed. This model can be used with any other model. Small-scale mockups【模型】 are created followed by iterative modifications until the final prototype is developed. The software developer can get valuable feedback from the users throughout the development process of the project. This user involvement results in increased user acceptance of the final implementation. However, this model can take excessive time in developing the prototypes. The concept of prototyping model is shown in Figure 10.12.

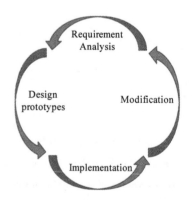

Figure 10.12　The prototyping model

10.3.4　Spiral model

In the spiral model, there are spiral shaped iterative processes. In each iteration of the development process, risks are identified and then resolved. Software in the spiral model is developed by repeated iterations of the following phases.

（1）Determine objectives【决定目标】
（2）Identify and resolve risks【识别与化解风险】
（3）Developments and Test【开发与测试】
（4）Plan the next iteration【计划下一次迭代】

The concept of prototyping model is shown in Figure 10.13.

Figure 10.13　The spiral model

10.3.5　Agile model

In the agile model, software development is iterative and incremental, where requirements and solutions evolve through interaction between customers and project teams. The Manifesto for Agile Software Development introduced the agile model in the late 1980s. A set of guiding concepts for the implementation of agile projects are referred to as Agile Principles. There are 12 principles for the agile model.

（1）Satisfy end-users【满足客户】 through early and continuous delivery of software.
（2）Welcome requirement change【需求变更】, even late in development process.
（3）Deliver working software more frequently.
（4）End users and developers work together throughout the project development process.
（5）Provide good environment to the project developers and trust them.
（6）Face-to-face communication is the most effective and efficient form of conveying information.
（7）Working software is the fundamental measure of progress.
（8）The stakeholders of the project should maintain sustainable development indefinitely.
（9）Agility is enhanced by paying continuous attention to technical excellence and good design.
（10）Simplicity is important.
（11）Best architectures, requirements, and designs are developed by self-organizing teams【自组织团队】.
（12）The team adjusts at regular intervals for becoming more effective【为了变得更有效率，团队会定期的休息以作调整】.

Agile model produces high quality software in minimum possible time. However, need special expertise in the project team. Furthermore, customers unable to express their needs can be an issue.

10.4　CMMI and Software Process Management

10.4.1　CMMI Overview

CMMI (Capability Maturity Model Integration) is a process improvement framework that provides organizations with the essential elements of effective processes to improve their performance. It identifies software's process weaknesses and strengths and makes necessary changes in the process in order to transform its weaknesses into strengths. CMMI lists what activities, and processes are required to be done and not how these activities and processes are done.

Software Engineering Institute developed the initial CMM (Capability Maturity Model) in 1990 and it specifies software process maturity.

The CMMI Project was developed to integrate the different CMMs into a set of integrated models. The CMMI defines five different levels of maturity, as shown in Table 10.1.

Table 10.1 Five levels of CMMI maturity

Level	Process maturity basements	Feature
Level 1 (Initial)	Unpredictable results	A process is translated from ad hoc【特别的】 approaches, methods, notations, tools, and reactive management, etc
Level 2 (Managed)	Repeatable project performance	Project-level activities and practices are the primary focus
Level 3 (Defined)	Project performance	Focus is on crucial process areas to establish organization-level practices and activities
Level 4 (Managed)	Organizational performance	Results for level 3 projects can be used to make tradeoffs【折中】, with predictable results, among cost, quality, and timeliness
Level 5 (Optimized)	Reconfigurable organizational performance, and quantitative, process improvement	It is related with technology innovations【技术革新】 and process change management【过程变更管理】

10.4.2 Principles of modern (iterative) software process management

There are many principles of modern software process management, a few of them are discussed here:

（1）**Focus on the process architecture**【架构】: During software development, focus must be on architecturally important design decisions. A demonstrable balance is needed among the driving requirements【驱动需求】, architecturally significant design decisions【有架构意义的设计决策】, and lifecycle plans【生命周期计划】 prior to committing sufficient resources of the organization for full-scale development.

（2）**Confronts risks early by establishing iterative life cycles**【迭代的生命周期】: Instead of sequential process, an iterative process is required that refines understanding of the problem, plan, and its solution. Major risks need to be addressed early to increase predictability and to avoid expensive rework【修订】.

（3）**Emphasize component-based development**【基于组件的】: In order to reduce the amount of human-generated source code and custom development, project teams must move from a line-of-code【基于代码段的】 mentality to a component-based【基于组件的】 mentality within an existing architectural framework. A component is a cohesive set of pre-existing【预先存在的】 lines of code, either in source or in executable format, with a defined interface【接口】 and behavior【行为】.

（4）**Develop a change management environment**【变更管理环境】: In iterative development, project teams must work concurrently.

（5）**Use rigorous, model-based design notation**【基于模型的设计符号】: Models such as UML, supports semantically rich design notations. Instead of using traditional design approaches, visual modeling and machine-processable【可被机器处理的】 languages are more useful as they allow more objective assessment.

（6）**Instrument the process for assessment of quality control and progress:** The process should integrate progress assessment【阶段性评估】 and quality of all intermediate products by using well-defined measures.

（7）**Develop configurable**【可配置的】**, and economically scalable**【可伸缩的】 **process:** The process must make sure that it is economically scalable by using extensive process automation and common architecture patterns, and configurable to a broad spectrum of applications. The organization should make sure that all their projects employ a set of best practices for project management and context【上下文】 independent workflows【工作流程】, checkpoints【检查点】, metrics【矩阵】, and artifacts.

10.5　Importance of Documentation

Documentation is important for software development, because it properly guides the developers for the efficient maintenance of the system. Various types of documentation are offered for software development, such as user documentation 【用户文档】, system documentation【系统文档】 and technical documentation【技术文档】.

10.5.1　User documentation

The software development team usually offers a user manual【用户手册】 to the customers. It shows step-by-step procedure of using the software and provides a detailed tutorial for the users to guide them about all the features of the software.

10.5.2　System documentation

System documentation need to be written so that the software maintenance team can properly maintain the software. There are four types of system development documentations: Analysis phase, design phase, implementation phase and testing phase.

（1）In the analysis phase, the collected information is documented properly. In addition, the sources of information are defined and details of requirements and methods are stated with its rationale【基本原理】.

（2）In the design phase, the details of tools required or used are documented.

（3）In the implementation phase, it is important to document every module of the code with comments and descriptive headers.

（4）In the testing phase, the developers must carefully document all details. Each test applied to the final product must be mentioned along with its result. Even any wrong result, failure, or weakness should be documented.

10.5.3　Technical documentation

It includes documentation about the installation and the servicing details of the software system.

Installation documentation【安装文档】 is about how to install the software on each computer. Service documentation【服务文档】 is about maintaining and updating the system.

10.6　Service-Oriented Architecture

SOA (Service-oriented Architecture)【面向服务的体系结构】 is an architectural approach for creating efficient systems. SOA is independent of any vendor, product technology or trend. EDI (Electronic Data Interchange)【电子数据交换】, CORBA (Common Object Request Broker Architecture)【公共对象请求代理体系结构／通用对象请求代理体系结构】 and DCOM (Distributed Component Object Model)【分布式组件对象模型,分布式组件对象模式】 are typical examples of SOA.

The objective of SOA is trying to achieve loose coupling【松散耦合】between interacting software agents. There is no standard for the exact composition of SOA, but many industry sources share some common factors.

- Service contracts are standardized.
- Services maintain loosely coupled relationships.
- Services hide logic from the external world.
- Services are reusable.
- Services are autonomous.
- Services minimize resource consumption【资源消耗】.
- High quality services are preferable.
- Services are discoverable.
- Services must be location transparent【位置透明的】.

10.7　References and Recommended Readings

For more details about the subject discussed in this chapter, the following books are recommended.
- Braude E J. Software Engineering: An Object-Oriented Perspective. New York: Wiley, 2001.
- Dale N, Lewis J. Computer Science Illuminated. Burlington, MA: Jones & Barlett Learning, 2013.
- Forouzan B, Mosharraf F. Foundations of Computer Science. London: Cengage Learning EMEA, 2008.
- Gustafson D A. Theory and Problems of Software Engineering. New York: McGraw-Hill, 2002.
- Layton C L. Agile Project Management for Dummies. Hoboken, NJ: Wiley, 2012.
- Lethbridge T C, Laganiere R. Object-Oriented Software Engineering: Practical Software Using UMLand Java. New York: McGraw-Hill, 2001.
- Pressman R S, Maxim B R. Software Engineering: A Practitioner's Approach. New York: McGraw-Hill, 2016.

- Schach S R. Object-Oriented and Classical Software Engineering. New York, NY: McGraw-Hill, 2005.
- http://www.agilemanifesto.org.
- https://msdn.microsoft.com/en-us/library/bb833022.aspx.

10.8 Summary

- The software lifecycle is a fundamental concept in software engineering. Software, like many other products, goes through a cycle of repeating phases.
- There are four phases of the development process in the software life cycle: analysis, design, implementation, and testing. Several models have been used in relation to these phases. Some of the popular models are: the waterfall model, the prototyping model, the spiral model, the agile model, and the incremental model.
- The SDLC begins with the analysis phase. The result in this phase is the requirement specification document that shows the function and scope of the target software without specifying how it will be done. This phase can be either procedure-oriented analysis or object-oriented analysis.
- In procedure-oriented design, the whole project is divided into a set of procedures or modules. In object-oriented design, the design phase continues by elaborating the details of classes.
- Modularity means breaking a large project into smaller parts that can be understood and handled easily. Two issues are important when a system is divided into modules: coupling and cohesion. Coupling between modules in a software system must be minimized. Cohesion between modules in a software system should be maximized.
- Some of the attributes of software quality are operability, maintainability, and transferability.
- The objective of testing phase【测试阶段】 is to find errors in the software. There are two kinds of unit testing: white-box testing (or glass box testing) and black-box testing. White-box testing is program-based testing, and black-box testing is specification-based testing. White-box testing uses the internal structure of the software. Black-box testing only tests software's inputs and outputs and focuses on the results other than the processes without knowing what is inside it and how it works.
- Software development models are the various methodologies or processes that have been developed in order to achieve the project's objectives. There are several models for the development process. Some of the most important and common models are waterfall model, incremental model, prototyping model, spiral model, agile model, iterative model, V model, and rapid application development (RAD) model.
- CMMI (Capability Maturity Model Integration) is a process improvement framework that provides organizations with the essential elements of effective processes to improve their performance.
- Various types of documentation are offered for software development, such as user

documentation, system documentation and technical documentation.
- SOA (Service-oriented Architecture)【面向服务的体系结构】 is an architectural approach for creating efficient systems.

10.9　Practice Set

1. Define software development life cycle.
2. Name some of the models for the software development process.
3. What are the phases of software development life cycle?
4. What is the difference between coupling and cohesion?
5. Describe the five distinct levels of CMMI.
6. What are the principles of modern software management?
7. What are the phases of software development process?
8. What is meant by software crisis?
9. What are the goals of software engineering?
10. What is the purpose of testing phase? Distinguish between white-box testing and black-box testing.
11. What is agile model for software development? What are its principles?
12. Write the importance and types of documentation.
13. What is modularity?
14. Which phase is about defining the users, requirements, and methods?_____.
    a. Analysis　　　b. Design　　　c. Implementation　　d. Testing
15. Which phase of system development process includes writing the program?_____.
    a. Analysis　　　　　　　　　　b. Design
    c. implementation　　　　　　　d. Testing
16. Which type of testing is involved in testing a software system?_____.
    a. Black-box　　b. White-box　　c. Neither a nor b　　d. Both a and b
17. _____ must be minimized between modules in a software system.
    a. Coupling　　b. Cohesion　　c. neither a nor b　　d. both a and b
18. _____ must be maximized between modules in a software system.
    a. Coupling　　b. Cohesion　　c. neither a nor b　　d. both a and b
19. Draw a data flow diagram for food ordering system.
20. Draw data flow diagram for the assignment equation: SUM = A+B × C;
21. Draw entity-relationship diagram for the organization structure of your school.

Unit 11
Data and File Structures

In this chapter, we focus on two aspects. One is about data structure. What are the most common data structures? How to add or delete data from a specific data structure? The other is about file structure. How to organize records in a file? How to retrieve a record? And what if a collision happened when you hash a key to an address in a hashed file?

11.1 Abstract Data Types

A data type is composed of a type and a collection of operations to manipulate the type. For example, an integer variable represents a member of the integer data type. The realization of a data type is an abstract data type (ADT) 【抽象数据类型】, which is a software component. The interface of an ADT is defined according to a type and a set of operations on that type. While its inputs and outputs determine the behavior of each operation. An ADT does not specify how to implement the data type. The hiding of implementation details from the user and protecting it from outside access is referred to as encapsulation【封装】. The concept of ADT is an instance【实例】 of managing complexity through abstraction【复杂问题抽象化】. Complexity and its handling techniques are a central theme of computer science. When dealing with complexity, we assign a label to an assembly of objects or concepts and then manipulate the label in place of the assembly. Such a label is called a metaphor by cognitive psychologists. A particular label might be related to other labels or other pieces of information. A hierarchy of concepts and labels is formed by giving a label to this collection. This hierarchy of labels allows us to pay more attention on crucial issues while ignoring unnecessary details. Figure 11.1 shows us the model for an ADT.

Take a car as an example. Steering, accelerating and braking are necessary activities for its operation. On almost all passenger cars, the steering wheel is turned for steering, the gas pedal is pushed for accelerating, and the brake pedal is pushed for applying brakes. This design for cars can be regarded as an ADT with operations "steer", "accelerate", and "brake". These operations on two different cars might be implemented very differently; for example, with different types of engine, or front-wheel versus rear-wheel drive. Yet, most drivers know to operate many different cars because the ADT presents a uniform operating method. As a consequence, drivers are not required to have knowledge of the specifics of any particular engine or drive design. These differences are hidden deliberately from users.

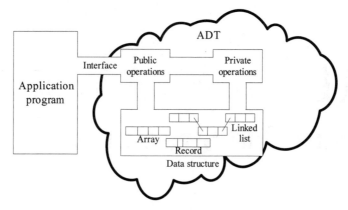

Figure 11.1 The model for an ADT

Data types are logical as well as physical. Logical data type can be defined in terms of ADT. Physical data type can be implemented as a data structure【数据结构】. Figure 11.2 illustrates the relationship between the logical and physical data types. The implementation of an ADT deals with the physical form of the associated data type, whereas the use of ADT elsewhere in the program is concerned with the associated data type's logical form. In an object-oriented programming language【面向对象编程语言】 such as C++, a class【类】 consists of an ADT and its implementation. A method or member function【成员函数】 implements operations associated with the ADT. Data members are the variables that define the space needed by a data item【数据项】. An object is an instance of a class【对象是类的实例】. It is created and allocated memory during the execution of a program.

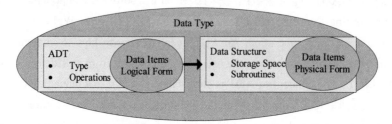

Figure 11.2 The relationship between data items, abstract data types, and data structures

11.2 List

A list【列表】 is defined as a finite, ordered sequence【序列】 of data items known as elements. Each element in the list has a position and a data type. Here we discuss a simple list implementation in which all elements have the same data type. The current number of elements in the list is referred to as its length. When length is zero, the list is said to be empty. The head is defined as the beginning of the list, and the tail is defined as the end of the list. Elements in sorted lists【有序列表】 are positioned in ascending order by value, while in unsorted lists【无序列表】, element values and positions have no particular relationship between them. Only unsorted lists are considered here. Array-based list【基于数组列表】 and linked list【基于链表列表】 are the two standard methods to implement lists.

In the array-based list, elements are stored in contiguous cells of the array. And this property must be maintained after insert, append【追加】and remove【删除】operations. It is easy to insert or remove elements at the tail of the list, so the append operation takes O(1) time. But if an element needs to be inserted at the head of the list, all elements must shift one position toward the tail to make room 【如果要在列头插入一个元素,那么列表中所有的元素都要向列尾方向移动一位】, as illustrated by Figure 11.3. This process takes O(*n*) time if the list has *n* current elements. If we want to insert an element at position *i* within a list having length equivalent to *n*, then *n* − *i* elements must move toward the end. Removing an element at position *i* requires moving *n* - *i* − 1 elements toward the head of the list.

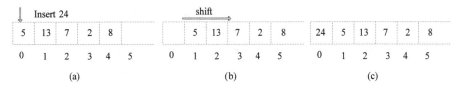

Figure 11.3　Inserting an element at the head of an array-based list

In the linked list, memory【内存】is allocated dynamically 【动态分配】when a new element is added into it. A series of nodes make up a linked list. A separate list can be made as a node class because nodes in a list are distinct objects 【独立的对象】(as opposed to simply a cell in an array). Figure 11.4(a) shows a linked list of four integers. The first node of the list is accessed from a pointer called head【头指针】. A pointer called tail【尾指针】is also kept to the last link of the list. In order to access to the end of the list faster and to allow the append operation to be performed in constant time 【常数时间】, another pointer called curr【当前指针】 is defined to indicate the position of the current element【当前元素】. The list length is stored explicitly and updated in any operation that alters the length of the list.

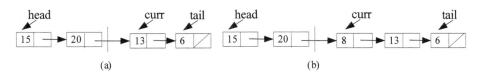

Figure 11.4　One possible implementation of a linked-list

In Figure 11.4(a), the curr pointer points directly to the current node. The logical position of the current element is shown by a vertical line between 20 and 13. If the we insert 8 between 20 and 13, the resulting linked list should be as shown in Figure 11.4(b). If *curr* is set to point to the preceding element 【前置元素】, adding a new element after the curr is much easier. Figure 11.5 shows the effect in this case.

Figure 11.6 shows the process of inserting new link node (having value 10) between the elements 23 and 12 of a link list. Figure 11.6(b) shows the linked list after insertion. ① shows the new linked node's element field 【数据域】. ② shows the new linked node's next field 【指针域】. It is pointing to the node used to be the current node (the node with value 12). ③ shows the next field of the node preceding the current position. It was pointing to the node having value 12; now it is pointing to the new node having value 10.

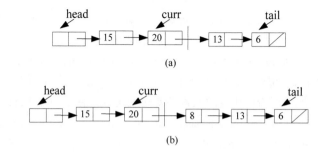

Figure 11.5　A better implementation of linked-list

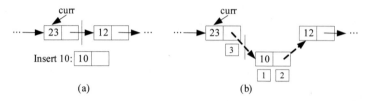

Figure 11.6　Node insertion in a linked list

Removing a node from the linked list needs only to redirect the appropriate pointer around the node to be deleted 【从链表中删除一个节点实际上只需要修改指向该节点的指针所指的地址】. Be careful to "leave" the memory【释放内存】for the deleted link node. The memory will be free by a call to delete 【通过一个系统调用，内存将被释放】. Figure 11.7 shows the remove method. In the linked list, inserting an element or removing an element takes O(1) time.

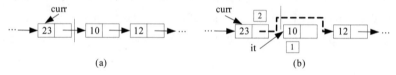

Figure 11.7　Node removal process in linked list

In Figure 11.7, ① shows "it" pointing the list node with value 10 being removed. ② shows the next field of the preceding list node, which is pointing the node following the one being removed.

Array-based list has the drawback that its length must be determined before the array can be allocated space, and its length cannot exceed beyond the pre-allocated size. Linked list has a benefit that it only needs space for the items actually on it, and its length can be exceeded as long as there is free storage available. A linked list is a very efficient data structure for sorted list, especially in insert and delete. It is also a dynamic data structure. The list can be empty at first, and then grow when new nodes are added. A node can be easily deleted without moving other nodes.

Array-based list has the benefit that there is no wastage of storage space if the list is completely filled. Linked list has the drawback that it needs an extra pointer to be added to all the list nodes 【基于链表列表的缺点是需要为所有节点增加一个额外的指针】. The overhead 【开销】 for links can be great if the size of linked list is large.

In general, if the length of list is small, it needs less space. Moreover, the linked list is more space efficient if its length is unknown. Conversely, the array-based list is more space efficient if the array is close to full and its size is predetermined.

11.3 Stack

The stack【栈】 is a list-like 【与列表相像的】 data structure. Elements can only be inserted or removed from one end in stack. For this reason, stacks are not as flexible as lists. However, in the applications that require only the limited form of insert and remove operations, stacks are more efficient than lists.

Long before the invention of the computer, stacks were used by accountants. They named the stack a "LIFO" 【后进先出】list, which means "Last-In, First-Out". Stacks remove elements in reverse order of their arrival. The element we access each time is called the top element. When a new element is added, we say it is pushed onto the stack. When removed, an element is said to be popped from the stack. Figure 11.8 shows a simple representation of stack runtime with push and pop operations.

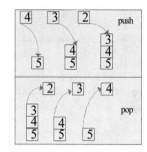

Figure 11.8 Simple representation of a stack with push and pop operations

The two common approaches of stack implementation presented here are array-based and linked stacks, which are analogous to array-based and linked lists, respectively.

The application of stacks in a most common computer application is perhaps not even visible to its users. Like the use of stacks during runtime of a subroutine call【子程序调用】 in most of the programming languages. The implementation of a subroutine includes pushing its information including return address【返回地址】, parameters, and local variables【局部变量】, onto a stack. The subroutine's information is referred to as an activation record. If a subroutine call contains another subroutine call, its activation record is also pushed onto the stack. The top activation record is popped off the stack on each return from a subroutine.

11.4 Queue

Similar to stack, the queue【队列】 is a list-like data structure that gives limited access to its elements. A queue has two kinds of operations. One is enqueue【入队】 operation, and the other is dequeue【出队】 operation. An enqueue operation allows elements only to be inserted at the back of a queue and a dequeue operation allows elements only to be removed from the front. Queues operate like standing in line at a subway station. If nobody cheats, then newcomers go to the back of the line. The person at the front of the line is the next to enter the subway. Thus, in queues, the elements are retrieved in order of arrival【在队列里，元素按照到达队列的顺序依次被访问】. The term queue has been in use by accountants long before the existence of computers. Queue is also called a "FIFO" 【先进先出】list, which means "First-In, First-Out". Figure 11.9 shows the representation of a queue.

There are two implementations of queue: the array-based queue, and linked queue. In practice, a queue can also be treated as a circle【在实际中，队列也可以是循环队列】. With the help of a modulus operator【取模操作】 (denoted by % in C++), a queue can circularly direct from the highest-numbered position to the lowest-numbered position in the array. In this way, positions in the array are numbered from 0 to size-1, and position size-1 is defined to immediately precede position 0 (which is equivalent to position size % size). Figure 11.10 illustrates this solution.

In Figure 11.10, (a) represents the queue after the insertion of initial four numbers 20, 5, 12, and 17. (b) represents the queue after the deletion of elements 20 and 5, following the insertion of 3, 30, and 4.

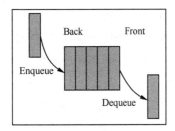

Figure 11.9 The representation of a queue

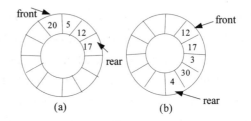

Figure 11.10 The circular queue with array positions increasing in the clockwise direction

11.5 Tree and Graph

A tree【树】is a connected undirected graph with finite set of elements called nodes and finite set of branches that connect the nodes. The degree【度】of a node is defined as the number of branches associated with a node. The number of branches directed toward the node is called the node's indegree【入度】 while, the number of branches directed away from the node is called outdegree【出度】.

The root【根节点】 of a tree is the first node, if the tree is not empty. The indegree of the root is zero. Except the root, each node in a tree has an indegree of exactly one. However, outdegree of any node in a tree can be zero, one or more. A leaf【叶节点】 is any node with an outdegree of zero. A node that is not a root or a leaf is called an internal node【中间节点】.

If a node has successor nodes【后继节点】, that is, if it has outdegree of at least one, it is called a parent【父节点】. A node with a predecessor is called a child【子节点】. Child node's indegree is necessarily one. Nodes with the same parent are referred to as siblings【同级节点】. A path is a sequence of nodes in which each node is adjacent to the next one. Every node has a unique path from the root. Any node in the path from the root to a particular node is called an ancestor, and any node in a path below a parent node is called the descendent. A subtree【子树】 is any connected structure below the root. Figure 11.11 shows an example of a general tree.

The level【层数】 of a node is its distance from the

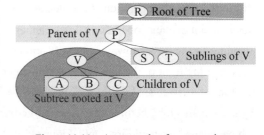

Figure 11.11 An example of an general tree

root. The root is at level zero and its child nodes are at level 1, and so on. The height, also called depth, of a tree is the highest level plus 1. So, according to definition, the height of an empty tree is -1.

A binary tree【二叉树】 is a rooted tree in which each node has no more than two children, that is the maximum outdegree of each node is two. Figure 11.12 shows an example of a binary tree.

The maximum height of binary tree having N nodes is N, whereas minimum height is $[\log_2 N] + 1$. Conversely, assume the height of a binary tree to be H, then the number of nodes in this binary tree ranges from H to $2^H - 1$.

A binary search tree (BST)【二叉查找树】 is a kind of binary tree that has a semantic property【语义特征】 on nodes' values. The value in each node must be greater than the value in any node in its left subtree and smaller than the value in any node in its right subtree 【每个节点上的值都大于左子树任何一个节点的值，同时，小于右子树任何一个节点的值】. Figure 11.13 shows an example of a binary search tree.

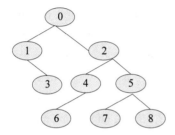

Figure 11.12　A binary tree

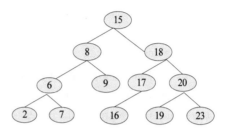

Figure 11.13　Example of a binary search tree

A tree can be stored in an array【树可以存储在数组中】. We assign numbers to the node positions in the entire binary tree【完全二叉树】, level by level, from left to right as shown in Figure 11.12. An array is an efficient data structure for storing a tree's data values, placing each data value in the array position corresponding to that node's position within the tree.

More about tree and binary tree data structure is available in reference materials listed at the end of the chapter.

In addition to list, stack, queue and tree, there is another very important data structure called graph【图】. Figure 11.14 shows an example of graph.

In a tree, a node is restricted to be pointed by at most one other node 【在"树"这种数据结构中，一个节点最多只有一个节点指向它】. If we remove this restriction on node, then we get a new data structure called a graph. A graph is made up of a group of nodes called vertices【顶点】 and a group of lines between the nodes called edges【边】. An undirected graph【无向图】 is a graph in which edges have no directions while edges in a directed graph 【有向图】 are directed from one node to another, as shown in Figure 11.14. Two vertices connected by an edge are called adjacent vertices【邻接点】. And a path【路径】 is a sequence of edges that connects two nodes in a graph.

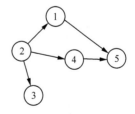

Figure 11.14　An example of graph

The most fundamental operation on graph is graph traversal【图的遍历】. It is defined as a process that accesses all of the vertices in a graph once and only once at a time starting from any vertex. Currently, there are two kinds of graph traversal methods, depth-first search【深度优先搜索】 and breadth-first search【广度优先搜索】. In a depth-first search process, we go to the deepest branch. When we have to backtrack, you take the branch closest to where you dead-ended. We back up as

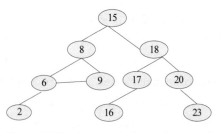

Figure 11.15 An example for graph traversal

little as possible in a depth-first search. Conversely, in a breadth-first search, we want to back up as far as possible to find a path originating from the earliest vertices. Both of the search methods can have more than two search paths. Take the graph in Figure11.15 as an example, starting from the vertices valued 15, in a depth-first search, one possible search path is 15→8→6→2→9→18→17→16→20→23. While in a breadth-first search, one possible search path is 15→8→18→6→9→17→20→2→16→23.

Here are examples of applications on graph. One application is the city map. A map of cities and the roads connecting the cities can be represented in a computer using an undirected graph. The cities are vertices and the undirected edges are the roads that connect them. If we want to show the distances between the cities, we can use weighted graphs, in which each edge has a weight that represents the distance between two cities connected by that edge. Another application of graph is in computer networks. The vertices can represent the nodes or hubs【中转站】, and the edges can represent the route【路由】. The cost of reaching from one hub to an adjacent hub can be defined by assigning weights on each edge. The router can find the shortest route between itself and the final destination of the data packet【数据包】 by the use of graph algorithms.

Except for the data structures mentioned above, there are hash table, red-black tree and B-tree. More about these three data structures is available in reference materials listed at the end of the chapter.

11.6 File Structure

Data is stored in the form of files in storage devices. A file is an external collection of related data treated as a unit【文件是相关数据的外部存储单元】. You may use file in the following situations.

- You need to store data permanently.
- The collection of data is often too large to store the entire data simultaneously in main memory【数据集太大以致无法同时在主存中完全加载】.

Files are stored in auxiliary or secondary storage【二级存储】 devices (disk). Files can be opened for both reading and writing. UNIX sees all devices as files. So, we can say that the keyboard is also a kind of file, although it cannot store data. For example, UNIX see all devices as files.

11.6.1 Access methods

When designing a file, the crucial issue is the way information in the file can be retrieved. An access method【访问方法】 defines how records can be retrieved: Sequentially or randomly.

Sequential access: In this access method, a file is accessed sequentially【文件被顺序访问】, and records are processed one after another, from beginning to end.

Random access: In this access method, records are accessed directly.

Different file structures use different access methods. Figure 11.16 shows the taxonomy of file structures.

Figure 11.16   Taxonomy of file structures

11.6.2   Sequential files

Records in a sequential file【顺序文件】 can only be read and written in sequence. To access a specific record within the file, the program must successively retrieve the previous records. An EOF (end-of-file) marker is put after the last record.

Sequential files are useful in applications which need to access all records in sequence from beginning to end. For example, if personal information about each employee in a company is stored in a file, you can use sequential access【顺序访问】 to retrieve each record at the end of month to print the paychecks. In this case, sequential access is more efficient and easier than random access【随机访问】. However, sequential files are not efficient for random access. For instance, let's assume that a bank has customer records that can only be retrieved sequentially. A customer who needs to withdraw money【取钱】 from an ATM would have to wait as the system checks each record from the beginning of the file until it reaches the customer's record. However, indexed files are very efficient for random access of large data files.

11.6.3   Indexed files

A key【键】 is the unique identifier 【唯一标识】 of data (or record) in the file which is made up of one or more fields【字段】. Sequential files may contain a key, but it is not used to locate the address of records. However, the address of the record should be known when accessing a record in a file randomly. An indexed file【索引文件】 contains records ordered by a uniquely identifiable key. For example, an account number might be the key for customer's record in a bank.

An indexed file is made up of a sequential data file and an index. The index itself is a very small file with only two fields: the key of the sequential file and the address of the corresponding record on the disk. Figure 11.17 shows the logical view 【逻辑视图】 of an indexed file.

Access methods for indexed files are sequential, random, or dynamic. Indexed file is efficient for random access. The address of some specific record can easily be identified according to the key. For example, the account number can identify the address of an account record in a file. But the indexed file needs an extra file (index), which is useless for sequential access.

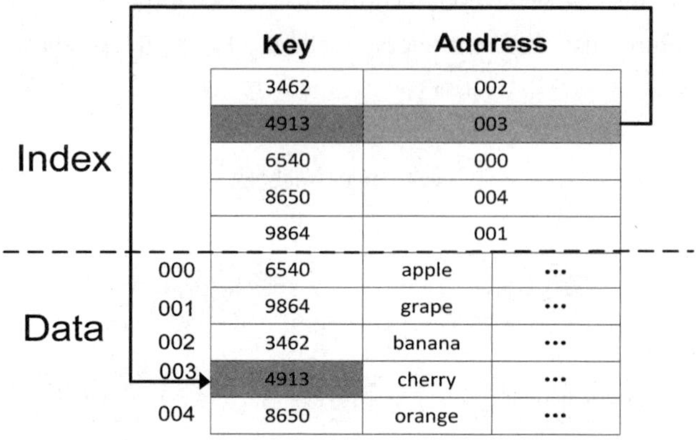

Figure 11.17   Logical view of an indexed file

An application of indexing is Internet search engines 【互联网搜索引擎】that use special indexing techniques 【专门的索引技术】for storing metadata 【元数据】 about web sites and content. Whenever a search is executed by a person, the search criteria entered on the screen 【输入到屏幕上的搜索条件】 is matched again these defined indexes. The information returned by the search is then ranked and displayed on the computer screen for review. In order to ensure that the contents returned by search engines are up-to-date and current, the search engines periodically 【定期】 update the indexed files.

Inverted file: One of the advantages of indexed file is that it can use alternate indexes 【替代索引】, each with a different key. For example, an employee file can be retrieved through employee department or last name rather than employee number. This type of indexed file is said to be an inverted file 【倒排索引文件】.

11.6.4 Hashed files

The index is used to map the key to the address in an indexed file. The hashed file【哈希文件】 does not need an extra file (index). It uses mathematical function to implement this mapping. The function returns the address once a key is given.

Many organizations use the process of hashing data into hash files for the encryption of their important data. Hash data is the representation of data in numerical form. A hash file uses mathematical function to convert it into a numerical string. A hash key decrypts this numerical string to make it understandable. A common implementation of hashing is in databases as a method of creating an index. Hash files are most often used as a method for verifying file size. This process is referred to as check-sum verification 【校对验证】. When a hash file is transmitted over a network, it must be divided into pieces and reassembled 【重组】 after it reaches its destination. In this case, the hash number represents file size. It can be used as a validating tool for the successful transmission of the entire file over the network. Other common use of hashing is in Internet routers, and for search mechanisms in databases.

The hashed file is also efficient for random access. In hashed file, we can immediately find the address of some specific record through the use of a function, thus eliminating the need for an index and all of its associated overhead. In contrast with indexed file, the hashed file can be space-efficient and quick to find the key of an address. However, hashed files sometimes may have their own problems.

1. Hashing methods

Hashing methods are used for key-address mapping【键-地址映射】. Some of the hashing methods are direct hashing, division hashing 【分治散列法】, digit extraction hashing, folding, midsquare, rotation, subtraction, and pseudorandom 【伪随机】 method.

Direct hashing: In direct hashing, the key is the address without any algorithmic manipulation. Since the keys are different from each other and the addresses are different, it guarantees that there are no synonyms or collisions【冲突】as with other hash methods. So, the file must have a record for every possible key. For example, it is unwise to use a social security number as the key because it has nine digits, which means a huge file with 999,999,999 records is needed, but in practice, less than 100 are used. An example of a case where direct hashing method can be used is the analysis of total monthly sales by days of month.

Modulo division hashing: In modulo division hashing【模分治散列法】, the key is divided by the file size and the remainder plus 1 is used as the hash address of the key. The list-size in the algorithm represents the number of elements in the file. Modulo division hashing and direct hashing method is represented in Figure 11.18. Collisions are possible as different keys may map into the same address. The algorithm produces fewer collisions if the list size is a prime number 【质数】instead of a composite number 【合数】.

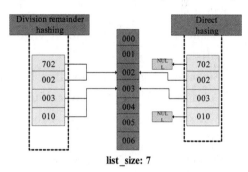

Figure 11.18   Modulo division hashing and direct hashing

2. Collision

A hash collision is a situation when a hashing algorithm produces an address for the key but that address is already occupied. Collision is an issue for the hashing methods. Except for direct method, all other methods may create a collision when you hash a key to an address.

Collision resolution: The collision of keys by addresses can be resolved by placing one of keys and data in another location. Some of the collision resolution methods are open addressing 【开放寻址法】, linked list resolution 【链表法】, bucket hashing 【桶散列法】, chaining 【链接法】, and quadratic probing 【二次探测】. Any hashing method can be implemented with any collision resolution method.

Open address resolution resolves collisions in the home area, that is, the area that contains all of the home addresses 【内部地址】. When a collision occurs, the addresses are searched for open or unoccupied record to store new data. In this address resolution technique, address of the item is not determined by its hash value. Data that cannot get space in the home address can be loaded in the next address (home address +1), which is a simple strategy. A major disadvantage to open addressing is that, the more collisions you resolve, the probability of future collisions also increases.

Linked list resolution resolve the issue of high probability of future collisions. In this method, the first record is loaded in the home address, but it has a pointer to the next record.

11.7 References and Recommended Readings

For more details about the subjects discussed in this chapter, the following are recommended:
- Dale N, Lewis J. Computer Science Illuminated. Sudbury, MA: Jones and Bartlett, 2004.
- Forouzan B, Gilberg R. Computer Science: A Structured Programming Approach Using C. Boston, MA: Course Technology, 2007.
- Forouzan B, Mosharraf F. Foundations of Computer Science. Cengage Learning EMEA, 2008.
- Gilberg R, Forouzan B. Data Structures: A Pseudocode Approach with C. Boston, MA: Course Technology, 2005.
- Goodrich M, Tamassia R: Data Structures and Algorithms in Java. New York: Wiley, 2005.
- Shaffer C A. Data Structures and Algorithm Analysis. Blacksburg, VA: Dover Publications, 2012.
- Shaffer C A. Data Structures & Algorithm Analysis in Java. New York, Dover Publications, 2011.
- Singh M, Garg D: Choosing Best Hashing Strategies and Hash Functions. IEEE International Advance Computing Conference, 2009.
- Weiss M A. Data Structures & Algorithm Analysis in C++. England, Pearson Education Limited, 2014.
- Zaqout F, Abbas M, Hosam A S. Indexed Sequential Access Method (ISAM): A Review of the Processing Files. UKSim European Symposium on Computer Modeling and Simulation, 2011.
- https://en.wikipedia.org/wiki/Hash_table.
- https://en.wikipedia.org/wiki/Red-Black_tree.
- https://en.wikipedia.org/wiki/B-tree.

11.8 Summary

- A data type is composed of a type and a collection of operations to manipulate the type. The realization of a data type is an abstract data type (ADT) which is a software component. An ADT does not specify how to implement the data type. The hiding of implementation details from the user and protecting it from outside access is referred to as encapsulation.
- A list is a finite, ordered sequence of data items known as elements.
- The array-based list and the linked list are the two standard implementing approaches.
- Compared to a list, a stack is a "LIFO" list and a queue is a "FIFO" list. So, there are many variations on the implementations of stack and queue.
- A binary search tree is a kind of binary tree that has a semantic property on nodes' values. The

value in each node must be greater than the value in any node in its left subtree and smaller than the value in any node in its right subtree.
- In a tree, a node is restricted to be pointed by at most one other node. If we remove this restriction on node, then we get a new data structure called a graph. A graph consists of a set of vertices and set of edges. Each edge is a connection between a pair of vertices.
- The most common graph traversal methods are depth-first traversal and breadth-first traversal.
- A file is a collection of related data treated as a unit.
- An access method defines how records can be retrieved: sequentially or randomly. Records in a sequential file can only be read and written in sequence. An indexed file is made up of a sequential data file and an index. The index is used to map the key to the address in an indexed file. The hashed file does not need an extra file (index). It uses mathematical function to implement this mapping. Hashing methods are used for key-address mapping.
- A hash collision is a situation when a hashing algorithm produces an address for the key but that address is already occupied. Collision is an issue for the hashing methods.
- A file can be organized in sequential file, indexed file and hashed file.
- Collision resolution methods find a new address for the hashed data that cannot be inserted.
- The most commonly used collision resolution methods are open addressing resolution and linked list resolution.

11.9 Practice Set

1. How is an element in an array different from an element in a linked list?
2. How are the elements of an array stored in memory?
3. What is the function of the pointer in a linked list?
4. What is an abstract data type? In an ADT, what is known and what is hidden?
5. What is stack? Give an example of stack implementation.
6. Define a tree. What is a binary search tree?
7. Show the process of inserting and removing nodes in a linked list.
8. Give an example of a graph to show breadth-first search and depth-first search.
9. List some applications of stacks and queues.
10. List some applications of binary trees. List the functions and types of files.
11. In a random or sequential access file, what is the function of the key and the address?
12. In direct hashing or modulo division hashing of a file, how is the key related to the address?
13. Give an example of a collision and then propose at least two kinds of collision resolution methods.
14. What is an inverted file?
15. How are indexing and hashing used for searching in Internet?
16. Which type of the following file can be accessed randomly? _____.
    a. Sequential file   b. Indexed file   c. Hashed filed.   d. b and c

17. If a record needs to be accessed _____, an indexed file may be the best choice.
    a. sequentially   b. randomly   c. in order   d. none of the above

18. In which kind of the following collision resolution method, you try to relocate data from location 23 to location 24 when needed? _____.
    a. Open addressing   b. Linked list
    c. Bucket hashing   d. a and b

19. A hash file uses a modulo division method with 41 as the divisor. Please address the following keys.
    a.14232   b.12560   c.13450   d.15341

20. Here is a file of size 411, please use modulo division to address the following keys. You can use open addressing and linked list resolution to resolve the address when collisions happen. And show the position with a figure.
    a.10278   b.08222   c.20553   d.17256

Unit 12 Databases

This chapter presents information about databases【数据库】 and their working. First, we discuss differences between data【数据】 and information【信息】. Second, we discuss how to store【存储】 data in databases. Third, we list some famous databases and present several advantages of databases. Fourth, we discuss how to use and manage data in the databases by a database management systems (DBMS)【数据库管理系统】. Last, we concentrate on three-level database architecture【数据库体系结构】, database history, database models【数据库模型】, database operations【数据库操作】 and big data【大数据】.

12.1 Introduction

Data is a mere collection of meaningless original records【记录】. Data requires processing, organization, and interpretation to become information【数据只有通过处理、组织和解释后才能成为信息】. Data needs to be stored persistently【持久地】. Then how do we store data? Traditionally, data storage used individual and unrelated files【文件】, sometimes called flat files【文本文件】. These flat files were used by individual application programs【应用程序】 and owned by different organizations. This type of flat file system causes many problems such as data redundancy【数据冗余】 and inconsistency【不一致性】. Database can be a perfect solution【解决方案】 to avoid these problems and to use data more easily, conveniently and efficiently.

A database is a collection【集合】 of related and logically coherent data used by application programs in an organization.

12.1.1 Advantages of databases

Today, why do we use database to store data? Compared with the flat file system【文件系统】 the database system【数据库系统】 has several advantages. Therefore, database system is widely adopted for storing data【数据库系统被广泛用来存储数据】. Some of the advantages of a database system are listed below.

1. Fully structured data【数据的完全结构化】

A database system stores not only the data but also the relations between data【数据库系统不仅存储数据而且也存储了数据之间的关系】. So, a database can guarantee【保证】 data integrity【数

完整性】 and easy operation【易操作性】.

2. Less redundancy【低冗余】

In a flat file system, files between two or more application programs cannot be shared【共享】easily, so there is a lot of redundancy. However, in a database system, we can greatly reduce redundancy through normalization【规格化处理】.

3. Inconsistency avoidance【高一致性】

If the same data are stored in multiple locations【如果同样的数据被存储在多个地方】, then any modification in the data is required to be made in all occurrences of that data【那么对于该数据的任何更改都需要在其所有出现的地方进行相应改变】. In a properly designed database system, data inconsistency is greatly minimized. One essential feature in database system is to guarantee data consistency.

4. Efficiency【高效率】

Database systems reduce data redundancy. That's why they are more efficient than flat file systems.

5. Data integrity【数据完整性】

Data integrity refers to the overall consistency, accuracy, and completeness of data【数据完整性指的是数据的整体一致性、精确性和自身的完整性】. In a database system it is easier to maintain【维护】data integrity, because data is stored in fewer places.

6. Confidentiality【机密性】

Database systems usually provide many security strategies【安全策略】 that guarantee data confidentiality.

12.1.2　Application areas of databases

Some of the representative application areas【应用领域】 of databases are enterprises, telecommunication【电信】, universities, banking and finance, computerized library systems【电子图书馆管理系统】, inventory systems【库存系统】, flight reservation systems【航班预订系统】, and content management systems【内容管理系统】, and so on.

12.2　Database Management Systems

A database management system (DBMS) is a computer software application that is located between the user and the operating system【操作系统】, as shown in Figure 12.1. We can use it to define, create and maintain a database【我们可以使用数据库管理系统定义、创建和维护数据库】. In a DBMS, there are usually five essential components【组件】 including hardware【硬件】, software【软件】, data【数据】, users【用户】 and procedures【程序】.

The DBMS performs many functions【功能】. Some of them are as follows.

（1）Data definition【数据定义】.

（2）Data organization, storage and management【数据组织、存储和管理】.

（3）Data manipulation【数据操纵】.

（4）Database transaction【事务】 and operation management.

（5）Database establishment【创建】 and maintenance.

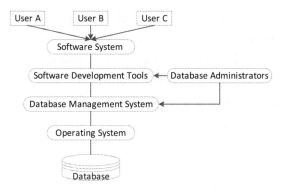

Figure 12.1　The position of the DBMS

12.3　Database Architecture

　　Modern database systems have a three-level architecture established by the American National Standards Institute Standards Planning and Requirements Committee (ANSI/SPARC)【美国国家标准协会/标准计划及需求委员会】. The three levels are internal【内模式/存储模式】, conceptual【概念模式/逻辑模式】, and external【外模式】.

　　The internal level defines an underlying description【底层描述】 and internal representation【内部表示】 of data in a database. It can also be referred to as the storage level. A database can have many external levels, but the internal level is unique 【一个数据库通常有很多外模式，但是内模式是唯一的】. It is concerned with low-level access methods【底层存取方式】, scalability【可扩展性】, cost【代价】, performance【性能】, and other operational matters. Therefore, the internal level directly interacts with the hardware【内模式是直接同硬件交互的】.

　　The conceptual level unifies various external views【外部视图】 into a compatible global view【全局视图】 and frees the users from dealing with the internal level directly. It defines the logical view【逻辑视图】 of the data, which is also called logical level. This level provides data model, programs that access the database, and the main functions of the DBMS. For the end-users, the conceptual layer provides an abstracted view of the database【对终端用户来说，概念模式提供了一种数据库的抽象试图】. This layer also defines queries【查询】.

　　The external level defines how the user views the organization of data in the database. A single database can have multiple views at the external level. The database along with its query processing languages resides at this level. It is responsible for defining relations on data, defining constraints【约束】, and changing the data coming from the conceptual level into a format that is familiar to the users【将来自概念模式的数据转化为用户熟悉的数据格式】.

12.4　The History of Database Systems

　　The history of database systems is shown in the Figure 12.2. Before 1960s, database applications were built on top of file systems. The data was difficult to access【访问】 or manage due to replication which creates redundancy and inconsistency【因为重复造成了数据的冗余和不一致】. Magnetic tapes【磁带】 were developed for data storage. Data could also be input from punched card decks【穿孔卡片组】 and output to printers【打印机】. However, tapes (and card decks) could be read only sequentially【磁带和穿孔卡片组只能顺序读取】. The history of database systems is shown in Figure 12.2.

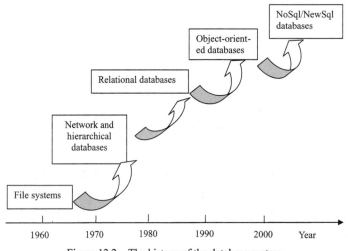

Figure 12.2　The history of the database system

In the late 1960s, with the widespread use of hard disks【硬盘】, it became easier to store and operate data. Therefore, network and hierarchical databases【网状/层次数据库】 were used extensively. However, people were not content with their complexity【复杂性】 and inconvenience.

In 1970s, more and more people were not content with the complexity【复杂性】 and inconvenience of network and hierarchical databases. Therefore, many researchers put forward the concept of relational model【关系模型的概念】. Relational databases【关系数据库】 became popular soon due to its simplicity and possibility of hiding of implementation details entirely【完全隐藏实现细节】 from the users.

In 1980s, IBM researchers developed techniques for the construction of an efficient relational database system【一种高效地创建关系数据库系统的技术】. Eventually, network and hierarchical databases were replaced by relational databases due to their ease of use【易用性】 and better performance. Relational databases freed the programmer from low-level implementation details. Research on parallel, distributed【并行/分布式数据库】, and object-oriented databases【面向对象数据库】 was also started in 1980s.

In early 1990s, SQL (Structured Query Language)【结构化查询语言】 language was developed to support querying and decision support systems. In late 1990s, with the explosive growth of the World Wide Web (WWW)【万维网】, research was started on database systems to support high transaction-processing【事务处理】rates with high reliability【可靠性】 and availability【可用性】.

In early 2000s, XML (Extensible Markup Language)【可扩展标记语言】 and the associated query language XQuery (XML Query), emerged as a new database technology. XML was designed to access collections of XML files like databases. This period also saw a significant growth in the use of open-source【开源】 database systems, particularly MySQL. In the latter part of the decade, several novel【新（颖）的】 distributed data-storage systems were built to handle the data management requirements of very large web sites【大型 Web 站点】 such as Baidu, Amazon, Facebook, Google, Microsoft and Yahoo. Eventually, with the continuous development of the applications and the unceasing change of demands, more and more people and enterprises start to recognize and use NoSQL (Not Only SQL) and NewSQL.

In early 2010s, with the growth of big data, web applications, and cloud computing【云计算】, high-performance data warehouses【高性能数据仓库】, data integration technologies【数据集成技术】, and Hadoop implementations were developed. The ever-growing shift to virtualization【虚拟化】, cloud computing, and semantic technologies【语义技术】 has shaped a new ICT landscape.

12.5 Database Model

A database model has its own logical structure and fundamentally determines the manner in which data can be stored, organized, and manipulated【数据库模型拥有自己的逻辑结构，并且从根本上定义了数据如何存储、组织和操纵的方式】. The main database models are hierarchical model【层次数据库模型】, network model【网状数据库模型】, relational model【关系数据库模型】, distributed model【分布式数据库模型】, object-oriented model【面向对象数据库模型】, semistructured model【半结构化数据库模型】, associative model【关联数据库模型】, Entity-Attribute-Value (EAV) model【实体—属性—值数据库模型】, context model【上下文数据库模型】, graph model【图数据库模型】, and multivalue model【多值数据库模型】 etc. The relational model is the most popular database model.

12.5.1 Hierarchical database model

The hierarchical database model organizes data into a tree structure【树形结构】. Every node【节点】 of the tree except the root【根节点】 has a parent node【父节点】. The nodes with the same parent are called twins or siblings【兄弟节点】. An example of a hierarchical model is shown in Figure 12.3.

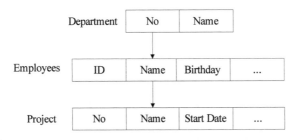

Figure 12.3 An example of a hierarchical model representing a department

12.5.2 Network database model

Network model permits many-to-many relationships【多对多关系】 in data that allows multiple parents. It is very similar to hierarchical model. In this model, the entities【实体】 are organized in a graph【图】 and data is organized in records and sets (Figure 12.4).

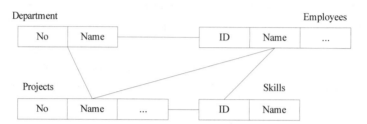

Figure 12.4　An example of a network model representing an enterprise

12.5.3　Relational database model

In the relational model, data is organized in two-dimensional tables【二维表】 called relations. We can operate data more easily and conveniently by using this model. Most of the database management systems (DBMS) are relational DBMS (RDBMS), such as MS SQL Server, Oracle, DB2, SYBASE, MySQL, SQLite, Realm, Microsoft Access, and so on. SQLite is a software library that implements a self-contained【独立的】, server-less【无服务器的】, zero-configuration【零配置的】, transactional SQL database engine【数据库引擎】. It is the widely deployed database engine in the world and it is a famous embedded database. Its features are simple and efficient. Its latest version is SQLite3. Realm is a high-performing and platform-independent【跨平台】 mobile database which is favored by many developers and hundreds of millions of users. Realm is faster, easier to use, and free. For the same operation, Realm usually needs less code lines than SQLite. Microsoft Access is a light-weighted small-scale database management system released by Microsoft, which is a member of the Microsoft Office Suite of applications. Table 12.1 provides the comparison of other mainstream databases. Figure 12.5 shows an example of a relational model.

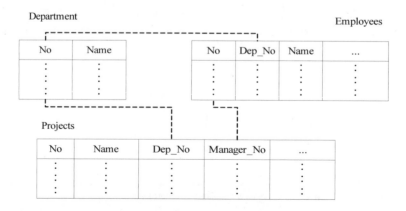

Figure 12.5　An example of a relational model representing an enterprise

Table 12.1 The comparison of common databases

Item/Database	MS SQL Server	Oracle	IBM DB2	MySQL
Vendor	Microsoft	Oracle	IBM	MySQL AB
Scalability	High	High	High	Medium
Performance	Variable	High	Variable	Variable
Cross-Platform【跨平台性】	No (Only for Windows)	Yes(Windows/Linux/Unix)	Yes(Windows/Linux/Unix)	Yes(Windows/Linux/Unix)
Complexity	Medium	High	High	Low
Open-Source	No	Yes	Yes	Yes
Security【安全性】	Low	High	High	Low
Connection-Method【连接方式】	ADO/DAO/OLEDB/ODBC	ODBC/JDBC/OCI	ODBC/JDBC	ODBC/JDBC/TCP-IP
Latest-Version【最新版本】	SQL Server 2014	Oracle 12c (12.1.2.0)	DB2 10.5	MySQL 5.7
Rank(Oct 2015, by DB-Engines)	3	1	6	2
Applications In	Microsoft/Huawei/China Telecom/China Mobile	Oracle/Baidu/Alibaba/China Mobile/Netease	IBM/Bank of China	Baidu/Alibaba/Meituan/Netease

ADO: ActiveX Data Objects【ActiveX 数据对象】; DAO: Data Access Object【数据访问对象】; OLEDB: Object Linking and Embedding Database【对象连接和嵌入数据库】; ODBC: Open Database Connectivity【开放数据库连接】; JDBC: Java Data Base Connectivity【Java 数据库连接】.

In addition to the famous databases mentioned in Table 12.1, some of the China-developed databases【国产数据库】 have been in use in recent years such as DM【达梦】, OpenBASE, OSCAR and KingbaseES. DM is the most widely used database in China. Table 12.2 represents a fact summary of several China-developed databases.

Table 12.2 A fact summary of several China-developed databases

Item/Database	DM	OpenBASE	OSCAR	KingbaseES
Vendor	Dameng Database【达梦数据库】	Neusoft【东软集团】	Beijing Shenzhou Aerospace Software Technology Company【北京神舟航天软件技术有限公司】	Kingbase【人大金仓】
Features	High performance; Highly reliable; Cross-platform; Easily usable; Low cost; Highly scalable	High compatibility; Multiple versions; Low cost; High speed	Efficient data processing ability; Multiple computing model; Complete application development interface	High speed; Highly reliable; Very stable; High security; High compatibility
Applications In	Public security department; Railway department; E-government affairs【电子政务】	Telecom; Hospital; Small and medium enterprises	Communication; Financial institution; Government sector	Education; Audit; Financial institution
Market Share	High	Medium	Low	Low

12.5.4 Distributed database model

The distributed database model is based on the relational model. In the distributed model, the database consists of multiple, interrelated databases physically distributed over different computer network sites【数据库是由多个相互关联并且物理上分布在不同计算机网络站点的数据库组成的】. To the end user, the distribution is transparent【透明的】, and the database acts as if it is centralized. A centralized distributed database management system (DDBMS) performs synchronization of data【数据同步】periodically and ensures that modifications performed on the data at one place will be automatically reflected in the data stored elsewhere. HBase and Google Bigtable are well-known distributed databases.

12.5.5 Graph database model

A graph database uses graph structures【图结构】for semantic queries【语义查询】with nodes, edges and properties to represent and store data. It allows even more general structure than a network database; any node can be connected to any other node. AllegroGraph, ArangoDB, Neo4j, and Bisty are several well-known graph databases.

12.5.6 Object-oriented database model

For the past decade or so, relational databases have been the industry standard, but new challenges are coming. Most of the applications applied object-oriented programming【面向对象编程】paradigm to RDBMS. This causes overhead of converting information between its representation in the database and the programming model. Object-oriented database【面向对象数据库】, also called Object Database Management System (ODBMS), aims to avoid this mismatch. ODBMS introduce key ideas of object oriented programming such as encapsulation【封装】, polymorphism【多态】, and inheritance【继承】.

12.5.7 Non-relational database model

A non-relational database【非关系数据库】is usually called NoSQL database. It became famous in the early twenty-first century to meet the needs of Web 2.0 companies and applications such as YouTube, Twitter, Facebook, Google, Amazon.com, Tencent (instant messaging service【即时通讯服务】in China), Baidu (search engine【搜索引擎】), Youku & Tudou (video sharing website), Taobao (e-commerce platform【电子商务平台】), Xiaonei (social network), Blogcn (famous blog website), and Sina (Chinese-language web portal). NoSQL is motivated by simplicity of design, better horizontal scaling to clusters【簇】of machines (which is a problem for relational databases), easy capture of all kinds of big data, faster data structure【数据结构】operations, persistent design【持久化设计】, interface diversity【接口多样性】, open source databases using clusters of cheap commodity servers【商用服务器】, finer control over availability, and use of BASE (Basically Available【基本可用】, Soft state【软状态】, and Eventual consistency【最终一致性】) between nodes. The data structures used by NoSQL databases make some operations faster in NoSQL (e.g. key-value【键值对】, wide column, graph, or document【文档】). NoSQL databases are increasingly used in big data, distributed computing【分布式计算】, cloud computing and real-time web applications【实时Web应用】. Based on the popularity ranking in October 2015, the most popular NoSQL databases are MongoDB, Apache Cassandra, Redis, and HBase.

Table 12.3 shows the comparison of several database models. Table 12.4 shows the comparison of common NoSQL databases.

Table 12.3 The comparison of several database models

Data Model	Type	Performance	Scalability	Flexibility	Complexity
Key-Value Store	NoSQL	High	High	High	None
Column-Oriented Store	NoSQL	High	High	Moderate	Low
Document-Oriented Store	NoSQL	High	Variable	High	Low
Graph Database	Graph	Variable	Variable	High	High
Relational Database	Relation	Variable	Variable	Low	Moderate

Table 12.4 The comparison of common NoSQL databases

Item/Databases	MongoDB	Cassandra	Redis	HBase
Language	C++	Java	C/C++	Java
Characters	Retain some good features of SQL	Support for large tables and Dynamo better	Run very fast	Support for billions of rows and millions of columns
License	AGPL	Apache	BSD	Apache
Database Model	Document store	Wide column store	Key-value store	Wide column store
Application Fields	Support dynamic querying or use indexes or need large memory	Write frequently	Update frequently and the size of databases can be predicted	Random and real-time access for large data
Rank (Oct 2015, by DB-Engines)	4	8	10	15

12.6 Relational Operations

How do we operate data in the database? These days, the relational database is still the most common type of database. So we mainly talk about operations on relations here.

A relational database is represented as a set of relations【一个关系数据库表示为一系列关系的集合】. Relational operations define what we can do with these relations. The main operations for a relational database are: insert【插入】, delete【删除】, select【查询】, update【更新】, project【投影】, join【连接】, union【并集】, intersection【交集】, set-difference【差集】, cartesian-product, and rename. We usually use the database query language SQL (Structured Query Language) to operate data in a relational database.

SQL is a special-purpose programming language designed for managing data stored in a RDBMS or for stream processing【流处理】 in a relational data stream management system (RDSMS). SQL is based on relational algebra【关系代数】 and tuple【元组】 relational calculus【关系演算】. It consists of a data definition language (DDL)【数据定义语言】, data manipulation language (DML)【数据操作语言】 and a data control language (DCL)【数据控制语言】. SQL became a standard of the American National Standards Institute (ANSI) in 1986, and of the International Organization for Standardization (ISO)【国际标准化组织】 in 1987.

The operations on database are shown in Table 12.5.

Table 12.5　The operations on database

Name	Function	Format
Insert	Insert a new tuple into the relation	Insert into Relation-name values (…)
Delete	Delete a tuple defined by a criterion from the relation	Delete from Relation-name Where …
Select	Search particular tuples from the database	Select from Relation-name Where …
Update	Change the value of some attributes of a tuple	Update Relation-name set attribute1 = value1,… Where …

Some of the relational operations are unary【一些关系操作是一元的】, i.e. they require a single relation as an operand【操作数】, while some operations are binary【二元的】, i.e. they take two relation operands. If we want to add some new records to a database, we need to use the insert operation. It is a unary operation on a relation. It is used to insert records into a relation. Figure 12.6 shows an example of insert operation.

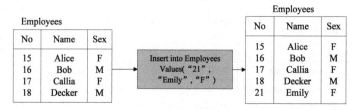

Figure 12.6　An example of insert operation

12.7　Databases for Big Data

With the rapid development of Internet technology, data is becoming richer and the amount of data is expanding sharply. We have entered the era of big data. Big data has several characteristics: volume【大容量】, variety【多样性】, velocity【高速度】, variability【可变性】and veracity【准确性】. So data representation, storage, management, sharing, visualization【可视化】, querying and processing are faced with new problems and challenges. However, traditional relational databases have some inherent disadvantages: poor expandability【难扩展】, slow velocity of reading and writing【读写速度慢】, high cost【成本高】and the limited capacity【容量有限】. So far, traditional database technologies have been incompetent for big data.

In order to overcome shortcomings in relational databases, new solutions and technologies are being proposed gradually. NoSQL is a solution. But it usually handles problems in some specific areas. For example, MongoDB is good at data storage based documents; Redis, TTServer and HandlerSocket have very high efficiency in reading and writing key-value data【键值数据】; AllegroGraph can store and process images better.

Clouds are also a solution. Cloud storage【云存储】 provides remote access【远程访问】 and storage of data, so cloud users can enjoy on-demand【按需预选】 high quality applications and services【服务】 without the burden of local data storage【本地数据存储】 and maintenance. Using cloud computing, users can get efficient computing power. Cloud computing has many advantages:

on-demand self-service【按需自助服务】, ubiquitous network access, location independent resource pooling, rapid resource elasticity, usage-based pricing and transference of risk【风险转移】. So many people say that cloud computing is expected to be the next generation information technology architecture【云计算有望成为下一代信息技术架构】. Many enterprises have their own cloud platform【云平台】, such as Google, Amazon, Baidu, Alibaba, and Inspur【浪潮】. In Baidu Cloud, a user can get 2 TB (terabytes)【太字节/百万兆字节】 free cloud room【免费云空间】 after registration【注册】. Cloud computing involves two major technologies: the parallel computing technology【并行计算技术】, such as MapReduce and the distributed storage technology【分布式存储技术】 of huge amounts of data, such as GFS (Google File System)【谷歌文件系统】 and HDFS (Hadoop Distributed File System)【Hadoop 分布式文件系统】. Apache Hadoop and Apache Spark are two popular products of Apache Hadoop, which is an open-source distributed data infrastructure for processing and storage of large data sets on computer clusters.

12.8 References and Recommended Readings

For more details about the subjects discussed in this chapter, the following books are recommended:

- Bahar H B. Database Management Systems Guidebook. CreateSpace, 2014.
- Beighley L. Head First SQL. O'Reilly Media, 2007.
- Blunt A D. Database System Concepts, United States. CreateSpace, 2014.
- Erl T, Puttini R , Mahmood Z. Cloud Computing: Concepts. Technology & Architecture, Prentice Hall, 2013.
- Forouzan B , Mosharraf F. Foundations of Computer Science. Cengage Learning EMEA, 2008.
- Johnson J L. Database: Models, Languages, Design. New York, Oxford University Press, 1997.
- Karau H, Kowinski A, Wendell P , Zaharia M. Learning Spark: Lightning-Fast Big Data Analysis. USA, O'Reilly, 2015.
- Kroenke D M , Auer D J. Database Concepts. Pearson, 2015.
- Lynch K S. Database Management Systems Intro. United States, CreateSpace, 2014.
- Marz N, Warren J. Big Data: Principles and best practices of scalable real-time data systems. Shelter Island: NY, Manning Publications, 2015.
- Mayer-Schonberger V , Cukier K. Big Data: A Revolution That Will Transform How We Live, Work and Think. New York, 2013.
- Peak K B. Database Systems: Design, Implementation & Management. CreateSpace, 2014.
- Silberschatz A, Korth H F , Sudarshan S. Database System Concepts. New York, NY: McGraw-Hill , 2010.
- Ullman J D , Widom J. A First Course in Database Systems, 2008.
- https://realm.io/

- http://www.bjsasc.com/
- http://www.dameng.com/
- http://www.inspur.com/
- http://www.kingbase.com.cn/kingbase/about/about-20.html.
- http://www.neusoft.com/cn/products&platform/0503/
- http://www.sqlite.org/index.html.

12.9 Summary

- A database is a collection of related and logically coherent data used by application programs in an organization.
- The database has many advantages, such as fully structural, less redundancy, inconsistency avoidance, efficiency, data integrity, and confidentiality.
- The DBMS has five essential components including hardware, software, data, users and procedures.
- The modern database systems have three-level architecture: internal, conceptual and external.
- The main database models are hierarchical model, network model, relational model, distributed model, object-oriented model, semistructured model, associative model, EAV model, context model, graph model, and multivalue model etc. The relational model is the most popular database model.
- In the relational database management system, data is organized in a set of relations.
- Structured Query Language is a programing language designed for relational database systems.
- Big data has several characteristics: volume, variety, velocity, variability and veracity.
- NoSQL and cloud computing are two solutions for handling big data.

12.10 Practice Set

1. Which five components does a DBMS contain?
2. Briefly explain the three levels of database and their relationships.
3. What is a database model?
4. What are the six database models? Which are popular today?
5. How are relations presented in the RDBMS?
6. What functions does a DBMS have?
7. List three common databases.
8. What characteristics does big data have?
9. What are the operations for a relational database?
10. What is the query language for the object-oriented database?
11. Each column in a relation is called _____.

　　　　a. an attribute　　b. tuple　　c. unoin　　d. record
　　12. The DBMS components such as computers and hard disks that allow physical access to data are known as the _____.
　　　　a. hardware　　b. software　　c. users　　d. application programs
　　13. _____ is concerned with low-level access methods, scalability, cost, performance and other operational matters.
　　　　a. External　　b. Conceptual　　c. Internal　　d. Physical
　　14. In three-level DBMS architecture, the layer that is closest to users or applications is the _____ level.
　　　　a. internal　　b. conceptual　　c. external　　d. hardware
　　15. The _____ database model arranges its data in the form of a graph.
　　　　a. hierarchical　　b. network　　c. relational　　d. distributed
　　16. The _____ level defines how the user views the organization of data in the database.
　　　　a. external　　b. conceptual　　c. internal　　d. physical
　　17. _____ is not a relational database.
　　　　a. SQL　　b. MySQL　　c. Oracle　　d. MongoDB
　　18. If you want to add a tuple to a relation, you use the _____ operation.
　　　　a. project　　b. insert　　c. update　　d. select
　　19. _____ is a unary operator.
　　　　a. Intersection　　b. Union　　c. Difference　　d. Select
　　20. You have relations Employees and Departments as shown in Figure 12.7. Show the resulting relation when you apply the following SQL statement: Select * from Employees where Dno = '001'

Employees

No	Name	Sex	Dno
15	Alice	F	001
16	Bob	M	001
17	Callia	F	002
18	Decker	M	003

Departments

Dno	Name
001	Sales Department
002	Planning Department
003	Human Resources Department

Figure 12.7　Exercise

Unit 13 Security

The following are some significant information leakage【信息泄露】 events:
- In June 2015, the United States Office of Personnel Management (OPM) was hacked. Sensitive personally identifiable information concerning more than 21 million people was stolen.
- In July 2015, the popular online cheating site Ashley Madison was hacked, and attackers hacked the information of around 37 million users. As a proof of the attack, the hackers posted online some of the hacked information.
- In July 2015, the Italian Surveillance Company Hacking Team was seriously hacked, and around 500 GB of client files, contracts, financial documents, and internal emails, were made publicly available for download.

Living in the information age, we are responsible for many aspects of our daily life that relate to the way our information is managed. From the above-mentioned hacking attacks, we learn that information is a valuable asset that needs to be secured from attacks. In this chapter, we will introduce some security threats, services and cryptography methods to protect and secure the information.

13.1 Security Goals

Information security【信息安全】 is the process of protecting sensitive information from unauthorized access 【未经许可的获取】, modification, inspection 【检阅，查看】, interception【拦截；窃听】 recording, use, disclosure【泄露】, destruction【破坏】, or disruption【扰乱；中断；瓦解】, to ensure its availability【可用性】, confidentiality【机密性】, and integrity【完整性】.

The traditional information security principles are confidentiality, integrity and availability, also known as the CIA triad【CIA 三元组】. The model is also sometimes known as the AIC triad (availability, integrity and confidentiality) to avoid confusion with the Central Intelligence Agency. The elements of the triad are considered the three most important components of security. Figure 13.1 shows the CIA triad of information security.

Figure 13.1 The CIA triad of information security

Confidentiality

In information security, confidentiality ensures that information is not made available or disclosed to unauthorized parties【在信息安全中，机密性保证信息不被未经授权的第三方获取或者披露】.

Integrity

In information security, data integrity【数据完整性】 defines the level of trust you can have in the information. This means that data can be modified only by appropriate mechanisms.

Availability

In information security, availability means that appropriate information must be available to authorized users when it is needed. Availability is ensured by rigorously maintaining the computing systems, protecting security controls and communication channels, maintaining correctly functioning operating system environment, providing adequate communication bandwidth, keeping necessary system upgrades, preventing data loss or interruptions in connections, and guarding against downtime or unreachable data.

13.2　Security Threats

13.2.1　Malware

Malware【恶意软件】, short for malicious software, is any software that is harmful to a computer user. Malware disrupt computer operations, compromise computer functions, gain access to private information, deny access to data and files, display unwanted advertising, or spy private information without user's knowledge or permission. Malware includes computer viruses【病毒】, worms【蠕虫】, Trojan horses【特洛伊木马】, bots, spyware【间谍软件】, adware, bugs, ransomware, rootkit, and other malicious programs. (Figure 13.2).

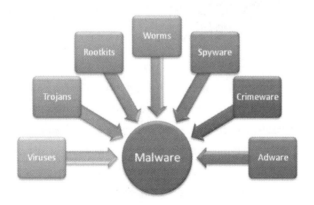

Figure 13.2　Common types of malware

In the following section, we describe the most common types of malware.

A virus is a program that is capable of embedding a copy of itself into another program【病毒是能够自我复制并且将自身嵌入其他程序中的一段程序】.

A worm is similar to virus in that it uses a network to replicate copies of itself onto other computers【蠕虫和病毒相似，它也是通过网络将自己复制到其他的电脑中】. In contrast to virus, worm does not need a host program to propagate. It is self-replicating and runs as a stand-alone program. A worm tends to cause problem on the networks, while a virus tends to cause problems on a particular computer.

In 1988, the first worm ever to be deployed on the Internet, named Morris worm【莫里斯蠕虫】, was released (Figure 13.3). It was named after its creator and releaser, Robert Tappan Morris. When the worm was first deployed, it infected over 6,000 computers in just a few hours.

Unlike the Morris worm that takes advantage of software vulnerabilities to spread, there exist other types of malware that rely on social engineering-based attacks to dupe users into downloading and installing it. The Trojan horse is one of them.

A Trojan horse, commonly known as a Trojan, is a program that appears to be helpful, but actually causes problems on the computer on which it is executing. A Trojan is difficult to track down because it disguises itself as a normal program or file to trick users into loading and executing it on their systems (Figure 13.4).

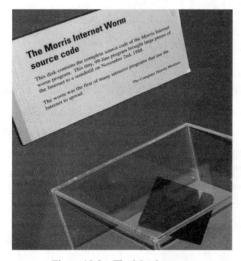

Figure 13.3   The Morris worm

Figure 13.4   Trojan horse

As the name implies, Spyware is software that aids in gathering information about a person or organization without their knowledge. It can capture information like web browsing habits, e-mail messages, usernames, passwords, and financial data. Its spying capabilities can also include interfering with network connections, and modifying security settings of browsers or software systems.

Bots are software programs created to automatically perform tasks and provide services or information that would otherwise be conducted as humans. Bots can launch DoS attacks, relay spam, gather information, render advertisements on websites, launch flood-type attacks against their targets, and distribute malware on downloading sites.

13.2.2   Security attacks

There are many types of computer security threats. In the following section, we describe several types of attacks.

Some attacks try to gain inappropriate access like password guessing attack【密码猜测攻击】. In this attack, password guessing is performed by repeatedly trying to log in to a system or application using different passwords. A computer program can attempt thousands of different passwords in one second. So it becomes possible to crack passwords. To address this problem, some authentication systems lock out accounts after a few failed attempts.

In a phishing attack【钓鱼攻击】, users may receive a deceptive email sent by the attacker, often containing links to malicious websites【恶意站点】. When users respond with the requested information, they often arrive at the spoofed site【欺骗的网站】 and enter their login credentials. The attacker later uses the logged credentials to log into the user's account and do whateve he wanted.

Spoofing【电子欺骗】 is a technique in which one person, program or an address successfully masquerades【乔装】as another by falsifying【篡改】 the data with the purpose of unauthorized access. This type of attack is may be an attempt to bypass【忽视】your firewall rules.

A buffer is an area of memory that often has a fixed maximum size【缓冲区是一个内存区域，通常有一个固定的最大值】. In buffer overflow attack【缓存溢出】, a program attempts to put more data in a buffer than it can hold, resulting in overrunning the buffer's boundary and overwriting the adjacent memory locations.

A denial of service (DoS)【拒绝服务攻击】 attack is designed to make a machine or network resource essentially useless to its intended users by flooding it with large quantities of external communication requests or useless traffic. The attacker sends many packets to the target website that keep the website too busy to service the legitimate users that are trying to access it【拒绝服务攻击的目的是使机器或网络资源本质上是无用的，它通过大量的外部通信请求或无效的网络，保持站点过于忙碌，导致合法用户无法获得服务】.The DoS attacks can cause unusual slow network performance, increasing unavailability of a particular website, increasing number of spam emails received, and inability to access data, website or other resources.

13.2.3 Cloud security threats

There are a lot of benefits of cloud computing driving a secular switch to the cloud. However, the shared nature of cloud computing creates the possibility of new security threats that can reduce the benefits made by the adoption to cloud computing. The key concern in the cloud that holds back cloud migration for some IT departments is the security of their data. Some of the biggest security threats in a cloud environment are given below.

Data breaches【数据泄露】: Cloud servers are an attractive target for attackers due to the vast amount of data stored on them. Cloud providers deploy controls for the security of virtual machines and hypervisors, but ultimately, cloud users are recommended to protect their own data against security breaches by using encryption and authentication mechanisms.

Ease of use【易用性】: The cloud can easily be used by malicious attackers for various purposes like malware distribution, spamming, password and hash cracking.

Data loss【数据丢失】: An intrusive or malicious attack can result in an intentional data loss. Permanent data loss in cloud data centers due to natural disasters or provider error can occur but are rare.

Malware infections【恶意软件感染】: Cyber criminals can use phishing attacks to deliver malware to targets through file sharing services.

Broken authentication: Weak authentication, keys, or passwords can result in data breaches. Multifactor authentication systems should be employed to make it difficult for the attackers to target the data.

Insecure APIs and hacked interfaces: The application programming interfaces (APIs) 【应用程序接口】 and interfaces tend to be the most exposed part of the cloud services because they can be accessed from anywhere on the Internet. An intruder can manipulate the customer's data through these APIs, thus compromising customer's integrity and confidentiality.

Account hijacking 【账号劫持】: Loss of credentials and passwords, buffer overflow attacks, and phishing can result in loss of control over a user account on cloud. An attacker can get control over it and use it to compromise the confidentiality, integrity, and availability of cloud services, such as redirecting customers to an inappropriate site, and providing false responses to customers.

System vulnerabilities and bugs 【系统弱点和漏洞】: Poor multitenancy in clouds can offer new routes to the hacker through different access points.

DoS attacks: DoS attacks are very common. These attacks can cause inaccessibility of cloud services.

Shared nature of clouds, shared dangers 【云的共享特性和危险性】: Many users perceive more threats in public cloud 【公有云】 because of the shared infrastructure, platforms, and applications.

When an organization wants to switch to the cloud, they should be well-aware of the cloud security threats for a successful transition.

13.2.4 Mobile security threats impacting smartphones and tablets

The rapid rise of adoption of, and reliance on, smart devices has also increase the threats of cybercrimes 【网络犯罪】 including risks of financial malware, malicious apps, spyware, mobile banking Trojans, zombies, and ransomware.

Some of the several mobile security threats are given below.

- The stealing of online banking credentials from the smartphones or tablets, by mobile banking Trojans, is very common nowadays.
- Mobile web browsers are also a target for cybercrimes.
- The attacker can also manipulate a smartphone as a zombie machine.
- The attacker can get control over a smartphone without the knowledge of its owner, and then the compromised phone can be used to make phone calls or record conversations.
- Smartphones and tablets can be compromised with malware by exploiting their resources.
- Attackers are now creating apps that block mobile security updates.
- Criminals can infect NFC-enabled devices (devices that have near-field communication capabilities), by gaining access to the victim's wallet accounts.
- RedBrowser, a java based Trojan, masquerades as a program and exploits smartphone's connections to social networks.
- Ghost Push, a malicious program on Android OS, cripples android-based smartphone resources

by automatically rooting the device, installing malicious apps directly to system partition, and then automatically unrooting it. This virus is very hard to detect.
- Mobile ransomware【勒索软件】attacks are growing since 2014, and have become most damaging scams on the Internet.
- SMS spam might be a form of phishing. The attackers use such messages to fools victims into giving up personal data.
- Scammers can capture personal data from the smartphones using free Wi-Fi hotspots.
- Apps from untrusted sources cannot be installed. For example, a malware named Android.Spyware.GoneSixty.Gen, discovered in 2011, uninstall itself within a minute after its installation. During this minute, it sends important data from the smartphone to a remote device. The user will never know it was ever there.
- An infected smartphone can be used to infiltrate nearby devices.
- Cryptocurrency【密码货币】 mining attack launches itself as a background service causing the phone's battery power to lose quickly or overheating.

Some of the security mechanisms to counter mobile threats are given below.
- Use of antivirus software and firewalls may protect the smart device and may prevent an intrusion.
- Biometric security can be used to prevent unauthorized access to the smartphone.
- Monitoring the battery consumption, battery usage, network traffic, and activities of various services can be helpful to detect certain malware applications.
- Regular backups of important data should be taken.
- Applications that request root access should be installed only by experts.
- During downloading or installation, users should be careful about the demands of the software.
- Bluetooth configuration should be set to non-discoverable mode by default.
- Wireless connection should be switched off when it's not in use.
- Sensitive data should not only be password-protected but it should also be well-encrypted.
- Setting up of SIM card lock or pattern lock can restrict access to the phone.
- Installation of third-party software【第三方软件】 should be disallowed.

13.2.5　IoT security threats

Security is extremely challenging due to the IoT environmental characteristics. With the growing number and variety of IoT devices, the potential for security threat is also escalating. Some of the security issues in IoT devices are: use of weak passwords on IoT devices; insecure web interfaces; flooding attacks; black holes; exhaustion of battery; lack of data encryption techniques【数据加密技术】; cloud control interfaces; jamming; tampering/forging, insufficient authorization; sniffing network traffic; and inadequate software protection etc.

Security issues may be resolved by training developers to integrate security solutions (e.g., firewalls【防火墙】, intrusion prevention systems【入侵防御系统】) into products and encouraging users to make use of IoT security features that are built into their devices.

13.3 Security Services

Standards have been defined for achieving security services and goals and for preventing security attacks.

13.3.1 Access control

When you log into a website, one of the most obvious security issues for you is to keep your accounts and information safe. Once providing your username and password, you are performing user authentication【用户认证】. Meanwhile, the username and password make up the authentication credentials【认证证书】. Authentication is the act of verifying someone's identity.

In the following section, four general types of authentication credentials will be introduced. The first type is based on something the user knows, such as username【用户名】and password【密码】. The second type is based on something the user filled in, such as CAPTCHA【验证码】. The third type is based on something the user has, such as smart cards【智能卡】. The fourth type is based on something the user is, in which the user's biometry is measured, such as fingerprint analysis【指纹分析】.

Password

Nowadays, passwords have been used for user authentication in most websites, operating systems, and other types of software. Although more sophisticated authentication mechanisms have been worked out, it is likely that password systems will be in use for some time (Figure 13.5).

A password is a secret word, phrase, or a string of characters that must be used for user authentication to prove identity and allow access to a resource. If the password is easy for the user to remember, then it means that it will also be easy for an attacker to guess. On the other hand, if the password is difficult for the user to remember, the user might need to write it down otherwise the password will need frequent resets.

Figure 13.5 Username and password

Guidelines used to improve the security of password include:
- Allow passwords of adequate length.
- Create a password that is easy for you to remember but difficult for others to guess.
- Don't use the same password across all of your online accounts.
- Don't tell your password to others.
- Don't use simple passwords that are easy to guess, such as the name of a pet, child, family member, etc.
- Use combinations of different kinds of characters, such as the combination of digits and letters.

CAPTCHA

A **CAPTCHA** (an acronym for "Completely Automated Public Turing test to tell Computers and Humans Apart") was coined in 2000 by Luis von Ahn, Manuel Blum, Nicholas Hopper and John Langford of Carnegie Mellon University.

A CAPTCHA is a challenge-response test most often placed within web pages to verify users as human. For example, humans can read text or figures from distorted pictures as the one shown in Figure 13.6, but current computer programs can't.

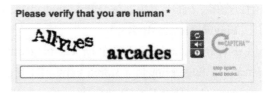

Figure 13.6 CAPTCHA

In what case do we need CAPTCHAs? CAPTCHAs have several applications for practical security, including (but not limited to):

- Preventing comment spam in blogs or forums: In order to truly guarantee that only human can enter comments on a blog, CAPTCHAs are needed.
- Preventing dictionary attacks【字典攻击】: In password systems, the users are required to solve a CAPTCHA after a particular number of unsuccessful login attempts to prevent dictionary attacks.
- Protecting website registration: By using a CAPTCHA, free services could be protected to prevent abuse by automated scripts.
- Worms and spam: CAPTCHAs offer a plausible way against email worms and spam by making sure that it is a human behind a computer.
- Search engine bots: CAPTHAs are required to ensure that bots won't enter a website.

Smart cards

A smart card (Figure 13.7) is tamper-resistant【防篡改】 card that has embedded integrated circuits 【集成电路】. The card can be programmed to self-destruct 【自毁】if a bad guy tries to gain access to the information stored on it. The memory, microprocessor, and other components that make up the "smart" part of the smart card are glued together in a way that it is not easy to take the card apart.

Smart cards are passive devices because they can function only if inserted into a reader integrated with a smart terminal, also referred to as CAD (Card Acceptance Device) or smart card reader. The CAD provides an electronic interface for communication, and power supply for the smartcard chip.

Figure 13.7 Smart card

The information stored in the card's memory and would only be accessible through the microprocessor.

A user can be authenticated by software which runs on a smart card's microprocessor. At the same time, it can guard any secret information stored on the card.

The smart card readers must be trusted when using smart cards for authentication. A rogue 【欺诈】 smart card reader that is installed by a bad guy can record a user's personal identification number, and if the bad guy can then gain possession of the smart card itself, he can authenticate himself to the smart card as if he were the user. In general, smart cards are hard to copy because of their tamper-resistance features.

Java-enabled smart cards, called Java Cards, have additional security features. It allows multiple applications to function independently and separately from each other on the same card【它允许多种应

用程序在同一张卡上独立且分开】.Nowadays, smart cards have cryptographic capabilities.

Fingerprint analysis

Fingerprint analysis is another technique used for user authorization. Fingerprint analysis compares a scanned fingerprint to a stored copy of the authorized user's fingerprint. It is considered a much stronger level of verification than username and password, and has become much more popular in recent years.

Nowadays, more and more products have fingerprint scanner hardware incorporated into the product itself (Figure 13.8). For example, a recent version of the Apple iPhone incorporates Touch ID (Figure 13.9), Apple's own fingerprint recognition technology. Some other fingerprint reader enabled smartphones are latest versions of Xiaomi, Lenovo, ZTE, Huawei, HTC, Samsung Galaxy, Meizu, Oppo, and OnePlus. For some other systems, the scanner is small but can be connected to a computer via a USB port.

Figure 13.8　Fingerprint Reader

Figure 13.9　Touch ID

13.3.2　Antivirus software

Antivirus software【杀毒软件】 is computer software designed to detect and remove malicious code or software. Nowadays, the most popular antivirus software Packages are: Rising【瑞星】, Norton【诺顿】, Kaspersky【卡巴斯基】, Kingsoft Antivirus【金山毒霸】, 360 Security【360 安全卫士】, Avira AntiVir【小红伞】, Grisoft AVG, and so on. Table 13.1 represents the comparison of various antivirus software packages.

Table 13.1　Comparison of various antivirus software packages

Antivirus Software	Performance	Characteristics	Antivirus Ability	Popularity
Rising	Full interception【全面监控】, Quick response【快速响应】	Start over before system program, Occupy a lot of resources	☆☆☆	☆☆☆
Norton	Two-way firewall【双向防火墙】, Heuristic technology【启发式技术】	Comprehensive information safeguard system, Intelligent virus analysis technology, Self-protective mechanism	☆☆☆☆☆	☆☆☆☆
Kaspersky	Heuristic analysis【启发式分析】, Anti-spam mechanism【反垃圾机制】	Run in background mode, Require computer with high standard configuration	☆☆☆☆☆	☆☆☆☆☆

续表

Antivirus Software	Performance	Characteristics	Antivirus Ability	Popularity
Kingsoft	Heuristic technology, Code analysis【代码分析】	Real-time monitoring, Check for viruses on compressed file【压缩文件】	☆☆☆☆	☆☆☆☆
360 Security	Fix weaknesses in systems, Privacy protection, Clean traces【清理痕迹】	360 Trojan firewall, 360 heuristic engine, QVM engine	☆☆☆☆	☆☆☆☆☆
Avira AntiVir	AntiVir【病毒防护】, AntiSpyware【间谍软件防护】, AntiRootkit【恶意软件防护】, AntiPhishing【链接扫描】	Easy to install, Occupy little resources, Less memory space	☆☆☆☆☆	☆☆☆
Grisoft AVG	Email Scanner【邮件扫描】, Community Protection Network, Real-Time Outbreak Detection	Advanced artificial intelligence on PC, Instantly converts every new threat	☆☆☆☆	☆☆☆

Traditional antivirus software uses signatures to identify malware. Substantially, when a malware arrives, it is analyzed by malware researchers. Then, once a malware is identified, a proper signature of the file is extracted and added to the signatures database of the antivirus software.

Some more sophisticated antivirus software uses heuristic analysis【启发式分析】 to potential malicious code. By either mutation【变种】 or refinements【改进】 by other attackers, many viruses can grow into dozens of slightly different strains, called variants. A heuristic approach 【启发式算法】 refers to the detection of more general patterns than the strict signature detection approach, so that it can hopefully remove multiple threats.

Antivirus software has some drawbacks, such as it can impact a computer's performance. What's more, as antivirus software itself usually runs at the highly trusted kernel level of the operating system, it creates a potential avenue of attack.

13.3.3　Security information and event management

Most of the successful computer attacks rarely resemble real attacks. This is why it is critical to view log files and correlate activities (that is, look patterns) across logs coming from various sources; they are often an effective way to detect attacks. Security information and event management (SIEM)【安全信息和事件管理】 system combines security information management (SIM) and security event management (SEM). A SEM system supports real-time analysis and correlates events, which can help the security analyst to take defensive actions. A SIM system collects data (such as event logs, system logs, transaction records) into a centralized logging repository for trend analysis and provides automated reporting for compliance and centralized reporting. SIM supports non-real-time data analysis. By bringing these two functions together, SIEM provide near real-time identification of security threats, supports forensic analysis of log records, and recovery of security events.

SIEM is not a security control or detection mechanism by itself, but it makes the already deployed security technologies and incident handling activities more efficient and effective. So, it is a great tool for any security team to detect incidents that would otherwise not be detected. SIEMs are available as either on-premises, or as public cloud-based services. A successful SIEM implementation has become a necessity for virtually every organization by playing a critical role in identifying patterns that indicate security breaches, detecting and handling incidents, and streamlining compliance reporting.

The leading SIEM products include IBM QRadar, HP ArcSight, McAfee ESM (Enterprise Security Manager), Splunk Enterprise Security, and LogRhythm. China Telecom also offers a fully managed SIEM service. China Pacific Insurance Group (CPIG) created a big data security management and control platform, dubbed Hawkeye, based upon Intel Security SIEM solution — McAfee ESM.

13.3.4 Security operations center

A security operations center (SOC)【安全运营中心】 is a centralized unit where enterprise information systems like websites, databases, data centers, servers, networks, applications, desktops, and other endpoints are monitored, assessed, and defended. A SOC is needed because a firewall and Intrusion Detection System (IDS)【入侵检测系统】 are not enough to detect and negate an incident before it can cause significant damage. A SOC is a team primarily composed of security analysts organized to detect, analyze, defend, prevent, report, and rapidly respond to cyber security incidents and threats. SOC can be outsourced or internal.

In order to bring complete protection to an organization from threats and risks, a SOC should be equipped with SIEM system along with threat intelligence, 24/7 security monitoring, and incidence response.

CloudSOC: It monitors cloud service use within an organization, enables security of organization's cloud apps, and audit application logs via SIEM systems, for example, IBM Radar, HP ArcSight, and so on.

13.4 Cryptography

Cryptography【密码学】 is the study of mathematically encoding and decoding messages.

Cryptography is an important component of achieving security goals. Careful implementation of cryptography in applications, along with well-designed and correctly deployed software, good policies and procedures, and physical security, can result in real security. In this section, we will discuss important cryptography concepts as well as some of the current cryptography approaches.

13.4.1 Symmetric-key cryptography

Symmetric-key cryptography【对称密钥加密】 is sometimes called private-key【私有密钥加密】, secret-key, single-key, shared-key, or one-key cryptography. It uses the same cryptographic keys for both encryption of plaintext and decryption【解密】 of ciphertext【密文】. It requires that both parties have access to the shared secret key.

In cryptography, encryption is the process of converting original form of message, referred to as plaintext into a form whose meaning is not obvious, called ciphertext. On the other hand, decryption is the reverse process. It is the process of decoding a message. A cipher【密码】 is an algorithm for performing a particular type of encryption or decryption, while the key to a cipher is the set of particular parameters that guide the algorithm.

The two types of symmetric-key cryptography algorithms are stream ciphers, and block ciphers. Stream ciphers【流密码】 encrypt the bits of a message individually. Block ciphers【分组密码】 take an entire block of plaintext bits simultaneously and encrypt them as a single unit. Block length of 64 bits (8 bytes) is commonly used in the DES (Data Encryption Standard) algorithm or 3DES (triple DES)【三重数据加密算法】 algorithm and block length of 128 bits (16 bytes) is commonly used in the AES (Advanced Encryption Standard) algorithm.

The two forms of encryption are substitution, and transposition. We first describe a strawman encryption【稻草人加密】 algorithm called a substitution cipher【替代密码】. Substitution cipher is an encryption algorithm that substitutes one character in the plaintext message with another character. The receiver performs the reverse substitution to decode the message. Caesar cipher【凯撒密码】, named after Julious Caesar, is one of the earliest known substitution cipher in which each letter in the plaintext is replaced by a letter certain number of positions (between 1 and 25) down the alphabet. For example, with a right shift of 3, C would be replaced by F, D would become G, and so on. Figure 13.10 relates the example of Caesar cipher.

The main drawback of symmetric key encryption is the requirement that both parties have access to the same secret key 【对称加密算法的主要缺点是需要双方都有同样的私密】. Here we present a few symmetric-key encryption algorithms.

Plain: ABCDEFG
a right shift of 23
Cipher: XYZABCD

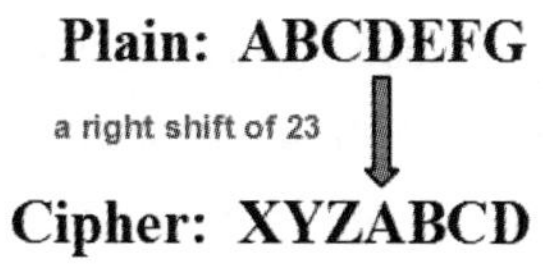

Figure 13.10 Example of Caesar cipher

Data Encryption Standard (DES)【数据加密标准】 is a predominant【主要的，卓越的】 symmetric-key algorithm that was adopted in 1977 by the National Institute of Standards and Technology (NIST)【美国国家标准技术研究所】. DES is a 64-bit block cipher, using a 56-bit key to customize transformation. There was a time, DES was probably America's most widely used symmetric encryption algorithm, not only in the financial area, but also in other industries as well. However, today, DES is quite vulnerable to brute force【暴力攻击】 attack (trying every possible key consecutively).

Triple DES applies the DES cipher algorithm three times and uses a different key for at least one of the three passes achieving a higher level of security.

Advanced Encryption Standard (AES)【高级加密标准】 is a version of the Rijndael algorithm designed by Vincent Rijmen and Joan Daemen. AES was adopted as a replacement for DES in 2001. AES supports key lengths of 128-bit, 192-bit or 256-bit, making it exponentially stronger than the 56-bit key size of DES.

13.4.2 Asymmetric-key cryptography

Asymmetric key cryptography【非对称密钥加密】 is sometimes called public-key encryption【公钥加密】. The issue with secret keys is their secure exchange over a large network or Internet. If the two parties who do not know each other want to communicate privately on the Internet, it would be extremely inconvenient for them to meet in person or talk over the phone to agree upon a key. The solution is asymmetric key cryptography which provides a way for them to do so without having to go to these lengths. In asymmetric key cryptography, each user has a pair of keys that are related mathematically. It is a complex relationship that a message encrypted with one key can be decrypted only with the corresponding partner key【这是一个复杂的关系，一条消息被根据一个钥匙加密，且仅可以用对应的伙伴的钥匙来解密】. What's more, one key is designed as the public key, which can be freely distributed, and the other key is the private key.

Here is an example. Suppose two users (Joan and Betty) want to securely communicate with each other. Each of them has their own public and private key pair. If Joan want to send a message to Betty, he needs to obtain Betty's public key first, which she makes readily available, and uses it to encrypt his message. Now no one can decrypt the message except Betty. Then Joan can send the message safely to Betty. Betty can decrypt it with her private key. In the same way, Betty can send a message to Joan only after encrypting it with Joan's public key.

Digital signatures【数字签名】 employ asymmetric key cryptography. A digital signature is a mathematical scheme for demonstrating the authenticity of a digital message or documents by appending extra data【数字签名是一种通过添加额外的数据来表明数字信息或文件的真实性的数学方案】.The goal of a digital signature scheme is to ensure that the sender of the communication is known to the receiver and the message being sent is non-repudiated. In digital signatures, the message is signed using the sender's private signing key, and then sent to the receiver, who can then use the sender's public key to verify the signature.

Here we present two representative asymmetric encryption algorithms, namely, RSA (Rivest-Shamir-Adleman) and ECC (Elliptic Curve Cryptography)【椭圆加密算法】.

RSA is the most influential public-key cryptosystems and is widely used for secure data transmission. The mathematical properties of the RSA algorithm are based on a simple number theory: Multiplying two large prime numbers【质数】 is very easy, but it is extremely difficult to factor【分解因子】 their large products. But with the rapid development of distributed computing【分布式计算】 and quantum computer theory【量子计算机理论】, the security of RSA encryption has been challenged.

ECC is another mathematical approach to build a public key cryptosystem and was originally proposed by Koblitz and Miller in 1985. As a new encryption method, elliptic curve encryption (ECC) algorithm has become a mainstream application gradually in e-commerce, smart cards, secure database, and so on.

13.4.3 Hash functions

Hash functions【散列函数】, also called one-way encryption【单向加密】 or message digests【信息摘要】, are algorithms that use no key. Instead, a fixed-length hash value【固定长度的散列值】 is computed based upon the plaintext. In this way, it is impossible to recover either the contents or length of the plaintext. Hash algorithms【散列算法】 are used to verify that the contents of the file have not been modified by a virus or an intruder. Hash algorithms are widely used today. The most common one is MD5 (Message Digest Algorithm)【消息摘要算法第五版】 and SHA1 (Secure Hash Algorithm 1)【安全散列算法1】.

MD5: MD5 is a widely used cryptographic hash function used for verification of data integrity through the creation of a 128-bit (16-byte) hash value. This function ensures that the information transmission is complete and consistent.

MD5 was designed by Professor Ronald L. Rivest of MIT in 1992. The purpose of MD5 was the secure compression of large files under a public key cryptosystem before being encrypted with a secret key. In 2004, Professor Wang Xiaoyun of Shandong University and her research team demonstrated collision attacks【碰撞攻击】 against MD5, SHA0 and other hash functions. A collision attack on a cryptographic hash occurs when two inputs result in the same hash value. In February 2005, Wang and her research team reported a method to find collisions in the SHA1 hash function. This method is being used in many security products. Previously, the time complexity of collision attack in SHA1 was 2^{80} operations, but this method reduced it to 2^{69}. In August 2005, Wang and her research team reported another method to find collision attack in SHA1 hash function, which further reduced the time complexity to 2^{63}, far less than original 2^{80}.

To crack the MD5 cryptographic algorithms, the computational power needs to be 2^{63}. Even with the fastest supercomputer【超级计算机】 today, it needs more than 100 years to crack it. But Wang Xiaoyun and her research team find a valid result within a few minutes with an ordinary PC. This finding once caused panic in some security departments.

An example of MD5 hashing can be seen in iOS' secret passwords, which avoid duplicate hashes if the password selected by two users is the same.

SHA1: SHA1 is a cryptographic hash function, designed by NSA (National Security Agency), along with NIST (National Institute of Standards and Technology) and was a US Federal Information Processing Standard.

SHA1 has a higher security, and robustness than MD5.

SHA2: It consists of two similar hash functions with block sizes 32-bit and 64-bit respectively.

SHA3: It was chosen in 2012. Its internal structure is different from other hash functions.

SHA2 and SHA3 have the same hash length.

13.4.4　Comparison of methods

Figure 13.11 shows three types of cryptography. Each one of the methods, symmetric-key cryptography, asymmetric-key cryptography, and hashing, has specific uses.

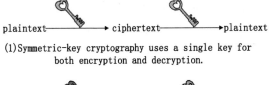
(1) Symmetric-key cryptography uses a single key for both encryption and decryption.

(2) Asymmetric-key cryptography uses two keys, one for encryption and the other for decryption.

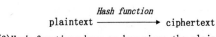
(3) Hash functions have no key since the plaintext is not recoverable from the ciphertext.

Figure 13.11　Three types of cryptography

Comparison of various encryption algorithms is shown in Table 13.2.

Table 13.2 Comparison of various encryption algorithms

Cryptographic Algorithms	Type	Speed	Key Length	Applications	Security
DES	Symmetric Algorithm	Fast	1 key of 56 bits	Banking, e-purse	★
Triple DES	Symmetric Algorithm	Fast	2 keys of 56 bits	Banking, e-purse	★★★
AES	Symmetric Algorithm	Flexible & Fast	1 key 128/192/256 bits	Banking, e-purse	★★★★
RSA	Asymmetric Algorithm	Slow	typical key: 1024 to 2048 bits	Banking, PKI, Digital Signature	★★★★
ECC	Asymmetric Algorithm	Slow	1 key up to 191 bits	E-Commerce, Smart Cards, Secure Databases	★★★★
MD5	Hash Function	Fast	1 key of 128 bits	File Transmission	Insecure
SHA-1	Hash Function	Fast	1 key of 160 bits	Security Applications, Protocols, Distributed Revision Control Systems	Insecure

13.4.5 Other cryptography methods

Quantum cryptography

Quantum cryptography【量子密码学】 is the science of existing knowledge of physics and quantum mechanics【量子力学】 to perform cryptographic tasks in a secure cryptosystem【密码系统】. It transmits a sequence of random bits on an optical network and also verifies if this sequence was intercepted or not【它在光网络上发送一个随机比特序列，同时验证该序列是否被截获】. The security model of quantum cryptography relies more on laws of quantum physics, rather than mathematics. The practical limitation of quantum cryptography is its necessary requirement of optical channel between the sender and the receiver【量子密码的实际限制是在发送者和接收者之间必要的光信道】.

It is considered to be a very secure cryptography method due to the following reasons.

- Messages transmitted using this method cannot be copied due to the unknown quantum state.
- Messages intercepted by an intruder will become useless because whenever an eavesdropper【偷听者】 tries to measure the state of the system, it will be disturbed.
- Messages intercepted by an intruder cannot be put back to their original state due to irreversibility【不可逆转性】 of the quantum property.

Steganography

Steganography【隐写术】 is the science of concealing information【隐藏信息】. Generally, the hidden messages appear to be something else: images, articles, shopping lists, or some other cover texts. In steganography, not only the contents of the message are concealed but also the fact that a secret message is being sent. Steganography includes the concealment【隐蔽】 of information within computer files. In digital steganography, electronic communications may include steganographic coding inside of a transport layer【传输层】, such as a document file, image file, program or protocol. Media files are ideal for steganographic transmission because of their large size data. For example, a sender might start with an innocuous【无害的】 image file and adjust the color of every 100th pixel to correspond to a letter in the alphabet, a change so subtle that someone not specifically looking for it is unlikely to notice it.

13.5 References and Recommended Readings

For more details about the subjects discussed in this chapter, the following books are recommended:
- Bishop M. Computer Security, Reading. MA: Addison Wesley, 2002.
- Catteddu D. Cloud Computing: Benefits, Risks and Recommendations for Information Security. Web Application Security. Springer Berlin Heidelberg, 72: 17, 2010.
- Deng C H, Hu C Y , Deng H F. ECC Solution of Security Problem with Medical Data Transmission in RFID Based Monitoring System. In Proc. of World Symposium on Computer Networks and Information Security (WSCNIS2014), pp. 55-59, June 13-15, Hammamet, Tunisia, 2014.
- Forouzan B. Cryptography and Network Security. New York: McGraw-Hill, 2007.
- John A, Nell. Computer Science Illuminated. Sixth Edition. Jones & Bartlett Learning: Lewis and Dale, 2014.
- Leyden J (August 19, 2005). SHA-1 compromised further. Crypto researchers point the way to feasible attack, The Register, retrieved 2016-04-14.
- Matoba O, Nomura T, Perez-Cabre E, Millan M S, Javidi B. Optical Techniques for Information Security. Proceedings of the IEEE, 97(6): 1128-1148, 2009.
- Randall J (March 11, 2005). Hash Function Update Due to Potential Weakness Found in SHA-1. RSA Laboratories, retrieved 2016-04-14.
- Whitman M E, Mattord H J. Principles of Information Security. Cengage Learning, 2015.
- Lowton S: A Guide to Security Information and Event Management. http://www.tomsitpro.com/articles/siem-solutions-guide,2-864.html. Retrieved 2016-07-10.
- https://web.archive.org/web/20150115033337/
- http://www.theepochtimes.com/news/7-1-11/50336.html.
- http://www.maigoo.com/maigoo/055soft_index.html.
- http://win8e.com/pc/zhishi/41820.html.

13.6 Summary

- We mentioned the CIA triad: confidentiality, integrity and availability.
- A virus is a program that embeds a copy of itself and inserts those copies into other programs. A worm is a type of virus that uses a network to copy itself onto other computers.
- To get the correct password, some attacks perform password guessing by repeated attempts of logging in to a system or application using different passwords.
- In a phishing attack, users may receive a deceptive email sent by the attacker, often containing links to malicious websites. A denial of service (DoS) attack is an attempt to make a machine or

- network resource essentially useless to its intended users.
- There are four general types of authentication credentials. The first type is based on something the user knows, such as username and password. The second type is based on something the user filled in, such as CAPTCHA. The third type is based on something the user has, such as smart cards. The fourth type is based on something the user is, in which the user's biometry is measured, such as finger print analysis.
- Antivirus software is computer software designed to detect and remove malicious code or software. Nowadays, the most popular antivirus software are: Rising【瑞星】, Norton【诺顿】, Kaspersky【卡巴斯基】, Kingsoft Antivirus【金山毒霸】, 360 Security【360安全卫士】, Avira AntiVir【小红伞】, Grisoft AVG, and so on.
- Security information and event management (SIEM) system provide real-time collection and analysis of security alerts generated by network hardware and applications. So, it is a great tool for any security team to detect incidents that would otherwise not be detected.
- A security operations center (SOC) is a centralized unit where enterprise information systems are monitored, assessed, and defended. In order to bring complete protection to an organization from threats and risks, a SOC should be equipped with SIEM system along with threat intelligence, 24/7 security monitoring, and incidence response.
- Cryptography is the study of mathematically encoding and decoding messages.
- Symmetric-key cryptography is sometimes called private-key, secret-key, single-key, shared-key, or one-key cryptography. It uses the same cryptographic keys for both encryption of plaintext and decryption of ciphertext.
- Asymmetric key cryptography is sometimes called public-key encryption. In asymmetric key cryptography, each user has a pair of keys that are related mathematically. It is a complex relationship that a message encrypted with one key can be decrypted only with the corresponding partner key. What's more, one key is designed as the public key, which can be freely distributed, and the other key is the private key.
- Hash functions, also called one-way encryption or message digests, are algorithms that use no key. Instead, a fixed-length hash value is computed based upon the plaintext. In this way, it is impossible to recover either the contents or length of the plaintext. Hash algorithms are used to verify that the contents of the file have not been modified by a virus or an intruder.
- Quantum cryptography is the science of existing knowledge of physics and quantum mechanics to perform cryptographic tasks in a secure cryptosystem. It transmits a sequence of random bits on an optical network and also verifies if this sequence was intercepted or not.
- Steganography is the science of concealing information within another file, message, image, or video.

13.7 Practice Set

1. Differentiate between cryptography and steganography.
2. What is CIA triad?

3. What is quantum cryptography?
4. Compare and contrast symmetric-key and asymmetric-key cryptography.
5. What are the three general approaches of presenting authorization credentials?
6. Distinguish between a worm and a virus.
7. What are cloud security threats?
8. What are the mobile security threats?
9. List 3 representative symmetric encryption algorithms.
10. List 2 representative asymmetric encryption algorithms.
11. List several representative antivirus software.
12. What are security information and event management (SIEM) systems and explain why they are important for security of an organization?
13. Make a brief description of the MD5 algorithm.
14. What is a digital signature?
15. What is encryption, and how does it relate to you as a student?
16. Create and describe the process for creating a password with a strong security level.
17. Which one is not the security attack? _____.
    a. password guessing   b. phishing
    c. DoS   d. fingerprint analysis
18. A DoS attack unusually cannot cause _____.
    a. slow network performance
    b. unavailability of a particular web site
    c. dramatic increase in the number of spam emails received
    d. better network performance
19. In symmetric-key cryptography, there is (are) _____ key(s).
    a. one secret   b. one private and one public
    c. either a or b   d. both a and b
20. In asymmetric-key cryptography, there is (are) only _____ key(s).
    a. one secret   b. one private and one public
    c. either a or b   d. both a and b
21. In symmetric-key cryptography, _____ possession of the secret key.
    a. only the sender has
    b. only the recipient has
    c. both the sender and the recipient have
    d. none of the above
22. In the digital signature method, the sender uses their _____ key to sign the message or digest.
    a. public   b. private   c. secret   d. none of the above

Unit 14
Theory of Computation

In chapters 1 through 13, we consider problem-solving as a core of computer science. In this chapter, we will see that we can answer questions such as: what kind of problems can be solved by a computer? Before execution of a program, can it be determined whether it will execute infinitely or it will terminate? How long does a program take to perform a specific task using a specific language? Theory of computation tries to answer all these questions. It is a scientific discipline that deals with how efficiently problems can be solved on a model of computation, using an algorithm and what are the limitations of computers.

14.1 The Turing Machine

The Turing machine【图灵机】 is an abstract machine【抽象机器】 devised by Alan Mathison Turing in 1936 to simulate mathematical operation, and is the foundation of modern computers【现代计算机】. Turing shows that a machine with the correct minimal set of operations can calculate anything that is computable, no matter what the complexity is.

Today, the common computer usually has a finite memory, but here we suppose that the Turing machine's memory is infinite【普通计算机内存有限，然而我们这里假设图灵机的内存是无限的】. In this section, we present a very simplified version【简化版】 of the machine to show how Turing machine works.

14.1.1 Informal description

The Turing machine mathematically models a machine that can operate on a tape【纸带】. The tape holds symbols, which the machine may read and write, one at a time using the tape head. The behavior is fully decided by a finite set of fundamental instructions【一系列有限的基本指令】 like "in state 50, if the symbol seen is 0, it will write 1; if the symbol seen is 1, it will become state 17; in this state, if the symbol seen is 0, it will write 1 and then become state 6" ,etc.

More precisely, a Turing machine contains: a tape, a controller【控制器】, and a read/write head【图灵机通常包括一条纸带、一个控制器和一个读/写磁头】. Its structure is shown in Figure 14.1. Someone has created a machine that embodied the classic look and feel of the machine presented in Turing's paper (Figure 14.2).

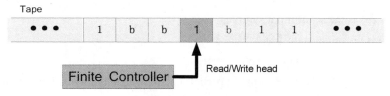

Figure 14.1　The Turing machine

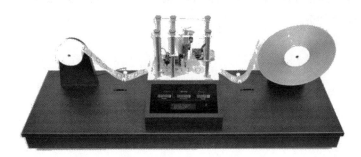

Figure 14.2　The Turing machine in the classic style

1. Tape

The tape contains several cells【小方格】, one next to the other. Every cell consists of a symbol from the finite alphabet【有限字母表】. For the sake of simplicity, we suppose that the machine here may accept only two symbols: a blank (b) and digit 1. The tape used here is freely extendable to the right and to the left【纸带能够左右自由的移动】, i.e., the Turing machine consists of infinite tape needed for its computation. We also suppose that the tape can deal with only positive integer【正整数】 data made up only of 1s. For instance, the integer 6 is represented as 111111 (six 1s). 0 is represented by the absence of 1s.

2. Read/Write head

The read/write head【读写头】 can read and write symbols on the tape and move to the left or to the right one (and only one) cell at a time. In some models【模型】the tape moves and the head is stationary【纸带移动而磁头静止】, whereas in some models, the head moves and the tape remains stationary【磁头移动而纸带静止】.

3. Controller

The controller【控制器】 functionally acts like the central processing unit (CPU)【中央处理单元】in modern computers. Actually, it is a finite state automaton【有限状态机】, a machine that has finite states and can move from one state to another depending on the input.

14.1.2　Formal description

Hopcroft and Ullman formally define a Turing machine as a 7-tuple【7个元组】:
M={Q, T, b, I, q_0, F, δ} where

- Q is a finite, non-empty set of states【非空状态集合】
- T is a finite, non-empty set of tape alphabet symbols【非空字母表/字符】
- B ∈ T is the blank symbol【空字符】(the only symbol allowed to occur on the tape infinitely often at any step during the computation)

- I ⊆ T - {b} is the set of input symbols
- q0 ∈ Q is the initial state【初始状态】
- F ⊆ Q is the set of final or accepting states
- δ = Q×T → Q×(T×{L, R, N}) is a partial function called the transition function【转移函数】, where L is left shift【左移】, R is right shift【右移】, and N is no shift.

Anything that operates according to these specifications is a Turing machine. The 7-tuple for the 4-state looks like this:

- Q = {A, B, C, D}
- T = {b, 1}
- I = {1}
- q_0 = A
- F = {D}
- δ=see transition table (Table 14.1) below

Table 14.1　Transition table for the Turing machine

Current state	Read	Write	Move	New state
C	b	1	L	D
C	1	1	N	A
B	b	b	R	C
B	1	1	L	D
A	b	1	L	B
A	1	b	R	A

We can create instructions【指令】 to put together the value of five columns in each row【每行】. For this elementary machine, we have only six instructions:

1. (C, b, 1, L, D)	2. (C, 1, 1, N, A)	3. (B, b, b, R, C)
4. (B, 1, 1, L, D)	5. (A, b, 1, L, B)	6. (A, 1, b, R, A)

The transition state【转移状态图】 diagram for the Turing machine is shown in Figure 14.3.

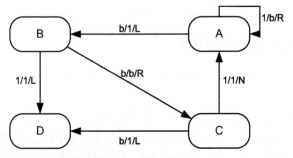

x/y/z: if x then write y, shift z

Figure 14.3　Transition state diagram for the Turing machine

14.1.3 Examples

1. Example 14.1

If a Turing machine with the above six instructions starts with the configuration shown in Figure 14.4, what will be the configuration of the machine after the execution of one of the above instructions?

The machine is in state B and the current symbol【当前字符】 is 1, which means that only the second instruction【第二条指令】, (B, 1, 1, L, D), can be carried out. The new configuration is also shown in Figure 14.4. After execution, the state of the controller【控制器的状态】 has been altered to D and the read/write head has moved one symbol to the left.

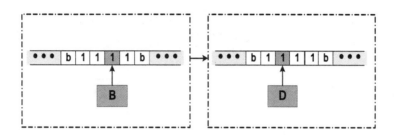

Figure 14.4   Example 14.1

2. Example 14.2

Figure 14.5 shows the working procedure【工作过程】 of the Turing machine for the **incr** (X)【增量】 statement【语句】 (set X=X+1). There are four states for the controller, S_1 to S_4. State S_1 is the starting state【开始状态】, state S_2 is the moving-right state【右移状态】, state S_3 is the moving-left state【左移状态】 and state S_4 is the halting state【终止状态】.

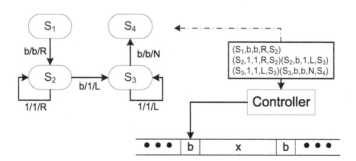

Figure 14.5   The Turing machine for the incr (X) statement

Figure 14.6 shows how the Turing machine can increment X when X = 2.

3. Example 14.3

Figure 14.7 shows the Turing machine for the decr (X) statement (set X=X-1).

Figure 14.8 shows how the Turing machine can decrement X when X = 2.

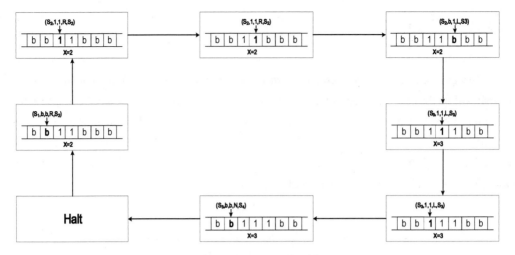

Figure 14.6   Example 14.2

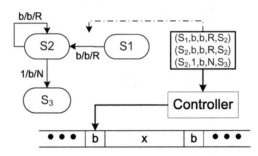

Figure 14.7   The Turing machine for the decr (X) statement

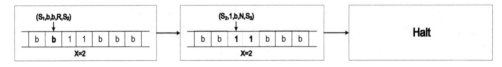

Figure 14.8   Example 14.3

14.1.4   The Church-Turing thesis

We have already proved that a Turing machine can implement some algorithms, such as decr (X) 【减量】 and incr (X). Can the Turing machine solve any problem that can be solved by a computer? We can find the answer in the Church-Turing thesis: If there is an algorithm that can do a symbol manipulation task 【符号操作任务】, then a Turing machine exists to do that task 【任务】.

14.2   Halting Problem

14.2.1   Introduction

The halting problem 【停机问题】 is a decision problem that computer programs run on a fixed

Turing-complete model of computation, i.e. all programs may be written by a given programming language which can use a Turing machine to do the same things. The problem is to decide, given a program and an input to the program, whether the program will halt or continue to run forever. Here we assume that we have enough memory【足够内存】and no time limitation【无时间限制】for the program's execution. It means before halting it may use arbitrarily long time【任意长的时间】and as much storage space【任意多的存储空间】. The question we discuss is whether the program will ever complete or run forever on a specific input【特定的输入】.

For instance, in pseudocode【伪代码】, like this: {while true: continue}, the program cannot stop; instead, it will run forever in an infinite loop【死循环/无限循环】. On the other hand, the program: {print "Hello, world!"} halts very quickly. A more complex program might be more difficult to analyze. Such program could be run for some fixed time to check if it halts【这种程序只能通过在固定的一段时间检查它是否能够停止】. If the program does not halt, there is in general no way to know if the program will eventually halt or run forever. Turing proved that there is no algorithm which can be applied to any arbitrary program and input to decide whether the program stops when run with that input.

14.2.2 Halting problem is not solvable

From the above we know that we can't write an algorithm【算法】that tests whether or not any program will terminate with a specific input. The computer scientist says "The halting problem is not solvable".

For the testing program, we will give a detailed proof【详细的证明】about its nonexistence. Our method, called proof by contradiction【反证法】, is often used in mathematics: first, we suppose that the program does exist, then we demonstrate that its existence will create a contradiction—therefore, it cannot exist. Here, we use three steps to prove this.

（1）For the first step, we suppose that a program, called Halting-solve, exists. Any program such as K can be regarded as input, and output is either 1 or 0. In theoretical computer science【理论计算机科学】, each program has an unsigned number【无符号数】and the program is written by a specific language【特定的编程语言】. The output of Halting-solve is 1 if the program K halts: The output of Halting-solve is 0 if the program K does not halt (Figure 14.9).

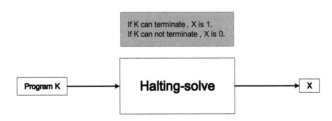

Figure 14.9 Step 1 in the proof

（2）In this step, we will use another program called Unusual that contains two parts: a copy of Halting-solve at the beginning and an empty loop【空循环】—a loop with an empty body—at the end. X is the testing variable【测试变量】for the loop and the output of the Halting-solve program.

（3）Having written the program Unusual, we use itself as test input. We can do this because we do not add any restrictions on program *K*. This is shown in Figure 14.10.

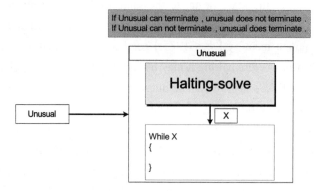

Figure 14.10   Step 3 in the proof

(4) Contradiction.

If we suppose that Halting-solve exist, we have the following contradictions:
Unusual does not terminate if Unusual terminates.
Unusual terminates if Unusual does not terminate.

From the above, we know that the Halting-solve program cannot exist and the halting problem is not solvable.

14.3   Solvable Problems

In computer science, we can say that, all problems can be usually classified into two types: solvable problems【可解问题】 and unsolvable problems【不可解问题】. The solvable problems also contains two categories: polynomial problems【多项式时间问题】 and non-polynomial problems【非多项式时间问题】(Figure 14.11).

There are a lot of problems which can't be solved by a computer. Halting problem is one of the most typical problems that cannot be solved by a computer. However, we are not interested in those problems. On the contrary, we usually want to care for the problems that can be solved by a computer. And we want to know the run time【运行时间】

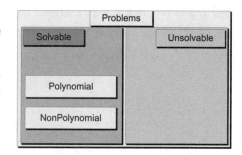

Figure 14.11   Taxonomy of problems

of a program, and the memory it needs. In the following section, we discuss the complexity of the solvable problems【我们将讨论可解问题的复杂度】.

14.3.1   Complexity of solvable problems

As we all know, if we want to know the complexity【复杂度】 of a solvable problem, we can compute the number of operations during execution of the program. But it is difficult to compute the run time exactly because of several reasons, for instance, the number of inputs, or the high speed of computers【高速运行的计算机】. In general, the big-O notation【大 O 记法】 is used to express the efficiency of an algorithm【表示算法的效率】. Here we will use the idea of this notation but we will not delve into its formal definition and calculation. In big-O notation, the number of operations is given as a function of the number of inputs. The notation $O(nlogn)$ means a program does $nlogn$ operations for n inputs, while the notation $O(n^3)$ means a program does n^3 operations for n inputs.

For instance, we have three algorithms to solve the same problem. The complexity of the algorithms are $O(n\log_{10} n)$, $O(n)$, and $O(n^3)$ respectively. We want to know how long does it take to run each of the algorithms on the same computer. Here we suppose n is 1,000,000 instructions per second. Figure 14.12 gives the answer.

```
1st program:   n = 1,000,000       O(nlog₁₀n) → 6*10⁶        Time → 6 sec
2nd program:   n = 1,000,000       O(n) → 1,000,000          Time → 1 sec
3rd program:   n = 1,000,000       O(n³) → 10¹⁸              Time → 31709 y
```

Figure 14.12　The execution time for different algorithms

14.3.2　Polynomial problems (P problems)

If we can find an algorithm that can solve the problem by a polynomial time【能够通过一个多项式时间解决的问题】, we call the problem is a polynomial problem. For instance, if an algorithm has a complexity of $O(\log n)$, $O(n)$, $O(n^2)$, $O(n^5)$, or $O(n^k)$, where k is a constant【常量】, we call it polynomial.

14.3.3　Non-polynomial problems (NP problems)

Some problems are deterministic【确定性的】 which can be solved through the formula step by step【能够通过方程式一步一步解决】. But the others are non-deterministic【非确定性的】 and there are no formulas or rules to solve the problems. We usually call these kind of problems as non-polynomial.

The travelling salesman problem (TSP)【旅行商问题】 is NP (Non-deterministic Polynomial) problem. Given a list of cities and their pairwise distances, the task is to find the shortest path【最短路径】 that visits each city exactly once and returns to the origin city.

14.4　References and Recommended Readings

For more details about the subjects discussed in this chapter, the following books are recommended:

- Forouzan B, Mosharraf F. Foundations of Computer Science. Thomson Learning, 2008.
- Hennie F. Introduction to Computability. Reading, MA: Addison Wesley, 1977.
- Hopcroft J E, Motwani R, Ullman J D. Introduction to Automata Theory, Languages, and Computation. Reading, MA: Addison Wesley, 2006.
- Kfoury A, Moll R, Michael A. A Programming Approach to Computability. New York: Springer,

1982.
- Minsky M. Computation: Finite and Infinite Machines. Engelwood Cliffs, NJ: Prentice-Hall, 1967.
- Sipser M. Introduction the Theory of Computation. Boston, MA: Course Technology, 2005.
- http://en.wikipedia.org/wiki/Turing_machine.
- http://aturingmachine.com/

14.5　Summary

- The Turing machine was designed to solve computable problems. It is the foundation of modern computers. A Turing machine contains: a tape, a controller, and a read/write head.
- Hopcroft and Ullman formally define a Turing machine as a 7-tuple. Anything that operates according to these specifications is a Turing machine.
- According to Church-Turing thesis: If there is an algorithm that can do a symbol manipulation task, then a Turing machine exists to do that task.
- A classical programming question is whether a program that can determine if another program halts can be constructed. Unfortunately, it has now been proved that this program cannot exist: the halting problem is not solvable.
- In computer science, problems can be divided into two categories: solvable problems and unsolvable problems. The solvable problems also contains two categories: polynomial and non-polynomial problems.
- Problems for which some algorithm can provide a solution in polynomial time is called polynomial problems. Non-polynomial problems can be solved by computers, but if the number of inputs is large, one could sit in front of the computer for months to see the result of a non-polynomial problem.

14.6　Practice Set

1. What are the components of the Turing machine and what is the function of each component?
2. When the head in Turing machine finishes reading and writing a symbol, what are the next steps?
3. Compare and contrast transition state diagram and transition table. Do they represent the same information?
4. What are the two categories of solvable problems?
5. Which operation will add 1 to the variable?
   a. Increment　　b. Decrement　　c. Loop　　d. Complement
6. The controller consists of _____ states.
   a. three　　　　　　　　　　b. four

c. a finite number of   d. an infinite number of

7. A _____ is a pictorial representation of the states and their relationships to each other.
   a. transition diagram   b. flowchart
   c. transition table   d. Turing machine

8. _____ is used to signify the program's complexity.
   a. the Turing number   b. big-O notation
   c. factorials   d. the Simple Language

9. The complexity of a problem is $\log_{10} n$ and the computer can carry out 1 million instructions per second. How long does the program take to execute if the number of inputs is 10,000?
   a. 1 microsecond   b. 2 microseconds
   c. 3 microseconds   d. 4 microseconds

10. After reading a symbol, the read/write head _____.
    a. moves to the left   b. moves to the right
    c. stays in place   d. any of above

11. Given a Tuing machine with a single instruction (A, b, b, R, B) and the tape configuration as shown in Figure 14.13, show the final configuration of the tape.

...	b	1	1	1	b

Figure 14.13   Exercises 10 to 11

12. Given a Turing machine with five instructions (A, b, b, R, B), (B, 1, #, R, B), (B, b, b, L, C), (C, #, 1, L, C), (C, b, b, R, B) and the tape configurationas shown in Figure 14.13, show the final configuration of the tape.

13. Display the sate diagram of a Turing machine that increments a nonnegative integer represented in the binary system. For instance, if the tape has $(101)_2$, it will be converted to $(110)_2$.

14. Show that the simulation of incr(X) in the Turing machine, as defined in this chapter, gives the correct answer when X = 0.

15. Display that the simulation of decr(X) in the Turing machine, as defined in this chapter show the result if X equals 0.

Unit 15
Artificial Intelligence

In fact, there is no universally-agreed【公认的，普遍认同的】 definition of Artificial Intelligence (AI). The following definition can be helpful to summarize this chapter. Artificial intelligence【人工智能】 is the study of programmed systems that can simulate, to some extent, human activities such as intelligently behaving, perceiving, reasoning, decision-making, problem-solving, thinking, learning, and acting【人工智能是这样的程序系统：它能在某种程度上模拟人的活动，例如智能行为、感知、推理、决策、解决问题、思考、学习和行动】.

15.1　Introduction

15.1.1　History of Artificial Intelligence

The roots of artificial intelligence can be traced back to more than 2400 years ago when the Greek philosopher Aristotle invented the concept of logical reasoning. In the 17th - 18th century, Leibniz and Newton made great efforts to finalize the language of logic. In the 19th century, George Boole developed Boolean algebra【布尔代数】 that laid the foundation of computer circuits. However, only in the 20th century, Alan Turing gave the main idea of a thinking machine by proposing the Turing test. In 1956, John McCarthy introduced the term "artificial intelligence". Although the history of artificial intelligence in the modern sense is not long, but after decades of development, its content becomes very rich.

15.1.2　The field of artificial intelligence

There are three schools【学派（专用于学术领域）】 of artificial intelligence: Symbolism【符号主义】, Connectionism【连接主义】, Actionism【行为主义】.

The Symbolism school【符号主义学派】 deems the human cognition【人类的认知】 primitive is symbols, and cognitive processes are symbol operation. The core problems of artificial intelligence are knowledge representation【知识表示】, knowledge reasoning【知识推理】, and application of knowledge. The Symbolism always tries to use mathematical logic to create a unified theory【统一的理论】 of artificial intelligence system, but always encounters the difficulties that some knowledge cannot be resolved.

The Connectionism school uses neurons【神经元】 as primitive objects, rather than symbolic processing. Connectionism thinks that the human brain is different from the computer, and proposed the Connectionism brain mode which replaced the symbolic manipulation of computer work. Connectionism argues that artificial intelligence should focus on structural analog【结构模拟】, that is to simulate human physiology based on neural network architecture【神经网络结构】, and agrees that function, structure and intelligent behavior【功能，结构和智能行为】 are closely related to each other. Different structures exhibit different functions and behavior. Based on artificial neural network structure, a large number of learning algorithms【（机器）学习算法】 have been proposed.

The behaviorism holds the opinion that intelligence depends on perception and action【感知和反应（行动，活动）】 (so it is called behaviorism). Behaviorism puts forward the perception-action mode of intelligent behavior, and suggests that intelligence does not require knowledge and reasoning. Artificial intelligence like human intelligence is just as evolutionary (so it is called evolutionary doctrine). Behaviorism claims intelligent behavior that only interacts with the surrounding environment can be manifested. Behaviorism thinks that Symbolism (Connectionism) operating mode of objective things is oversimplified in the real world, and thus it does not truly reflect the objective existence. Behaviorism also views that the study of artificial intelligence methods should use the behavior simulation method that the function, structure and intelligent behavior are inseparable. Different behaviors are the result of different functions and different control structures. The Behaviorism's methods are also doubted and criticized by other university schools that the behaviorist can only create intelligent insect behavior, but cannot create a human intelligent behavior.

The field of artificial intelligence is interdisciplinary and merges with a number of subjects such as psychology, computer science, neurology, philosophy, mathematics, linguistics, and electrical and mechanical engineering【心理学、计算机科学、神经病学、哲学、数学、语言学和电子与机械工程】.

15.2　Knowledge Representation and Expert Systems

15.2.1　Semantic networks

A semantic network【语义网络】 represents knowledge as a directed graph【有向图】. A directed graph is made of vertices (nodes), which are connected by edges (arcs).

In a semantic network, knowledge is represented as a set of concepts that are related to one another. For example, in Figure 15.1, the rectangle labeled "animal" defines the set of all animals, and "cat" defines the set of all cats which is a subset of the set animal.

In a semantic network, relations are shown by edges, which can define a subclass relation, an instance relation, and an attribute of an object (color, size,…) induced by the concept of different patterns or labels on the edges, as illustrated in Figure 15.1.

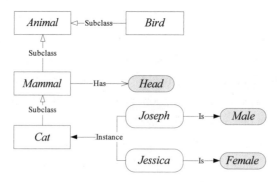

Figure 15.1　A directed graph to represent knowledge

15.2.2　Rule-based systems

A rule-based system【基于规则的系统】 uses rules for knowledge representation that can be used to deduce【演绎】 new facts from known facts. The rules express what is true if certain conditions are met. A rule-based database consists of rules expressed as a set of *if*... *then*... statements.

$$\text{If X then Y or X} \rightarrow \text{Y}$$

In the above expression, *X* and *Y* are named as the antecedent【前件】and consequent【后件】. Each rule is independent. The system mechanism is to scan the set of rules in a rule-based database, and choose one or several rules to execute one by one. After passing the conflict resolution strategy, the system can obtain the resulting actions.

15.2.3　Other representations

Knowledge can also be represented by using predicate logic【谓词逻辑】, non-classical logic【非经典逻辑】, and frame【框架】 etc. Among them, predicate logic is the most basic in logic systems. The logic itself has formed a complex classification system. Classical logic【经典逻辑】 includes propositional logic【命题逻辑】 and predicate logic【谓词逻辑】, while non-classic logic includes many kinds of logic which may have more than 2 values or may have the reasoning with uncertain condition, or may have the expression which contains more complex quantifiers【量词】 to represent richer meanings. A frame【框架】 is also a formal structure which is similar to a semantic network, such as a node in semantic network corresponds to an object in a frame. For more details, please refer to [2].

15.2.4　Extracting knowledge

Expert systems【专家系统】 use the knowledge representation languages discussed in the previous sections to perform tasks that simulate the decision-making ability of a human expert. They can be used in situations when that expertise is scarce, expensive or unavailable when needed. For example, in a medical diagnosis system, an expert system can be used to narrow down a set of symptoms【症状】 to a likely subset of causes, a task normally carried out by a doctor. It is built on the knowledge of a doctor specialized in the field for which the system is built.

The first step in building an expert system is to extract the knowledge from a human expert. This extracted knowledge is stored in a knowledge base; usually, it can be built as rule-based system【基于规

则的系统】.

A fact database【事实（实例）库】is needed in addition to the knowledge base for a knowledge representation language. The fact database in an expert system is case-based【基于案例的】, in which the collected facts are put into the system which is used by the inference engine. When a user applies a request, the system will make comprehensive use of the knowledge base and the fact base for reasoning【系统将结合知识库和实例库来进行推理】.

15.2.5 Typical contribution of the school of Symbolism

Although it is not so straightforward exciting as automatic programming, the machine proof of theorems may also be of greater significance in history. The proofs of mathematical theorem【数学定理的（机器）证明】 often need different methods, even for the same type of theorems.

Several centuries ago, R. Descartes and G.W. Leibniz proposed the idea of using the approach of computing in place of human work for the proof of mathematical theorems. In 20th century, David Hilbert, a mathematician, also studied this problem. Around 1950, A. Tarski, a Polish mathematician, proved that the Euclidean geometry theorem can be proved by calculation of computer, and gives the corresponding algorithm. However, even after many successors of improvement, and even with the high-speed computer to realize it, this algorithm can only prove Euclidean geometry theorem, and cannot prove other complex theorems.

From then on, this field had not made a breakthrough progress until Chinese scholar Professor Wu Wenjun【吴文俊】 proposed Wu's method in the late 1970s, which is about the mechanical theorem proving in elementary geometry. Using Wu's method, about 600 theorems have been proved by computer. This is of great significance for basic theory research.

15.3 Perception

All of intelligent techniques can be integrated into a body named as intelligent Agent 【代理；智能体】, which may be a robot, or an intelligent software, or other intelligent equipment etc. In a sense, it is a general name for a variety of intelligent objects. To intelligently respond to the input from outer sensors, an agent must be able to understand that input. That is, the agent must be able to perceive. In this section we explore the areas of research in perception【感知】that have been proved to be especially challenging: understanding images, sound, and language.

15.3.1 Image and sound processing

The basic techniques

Like human understanding of the world, the perception and cognition【认知】of intelligent system begins from the perception of image and sound. The perception of the image usually includes two steps. First is the identification of image features, named as image processing【图像处理】. Second is the analysis of the features, i.e. the results of image processing, to find out the intrinsic meaning. This process is named as image analysis【图像分析】.

Image processing has a large number of stages or activities. One is edge enhancement【边界增强】, which clarifies the boundaries between the image regions, so it is also an attempt to convert the photo image into a line graph. A large number of mathematical techniques can be used for it.

Region finding【区域发现】 is another image analysis activity, which identifies those areas in an image that have common features, such as color brightness, intensity, or texture. The common features may denote the areas belong to the same object, for example, the areas with the same background.

Smoothing【平滑】 is removing flaws and reducing noise within the image, which may also reduce the negative impact【负面影响】 and improve the quality of the image. However, smoothing has a negative impact that it may lose important information.

Image analysis is used to process an image into fundamental components in order to extract meaningful information and statistical data. Image analysis most often uses hypothetical method, which may guess the image object as an association component and then may try again in case of failure. This action is just as humans do. For example, sometimes we are very puzzled in a scene or image, but once we have a clue, we can easily identify it.

Speech recognition【语音识别】 technology is to convert the lexicon represented by human speech【由人类声音表达的词汇】 into computer-readable input, such as binary code or character sequence. However, in speech or voice recognition, speaker identification refers to identifying the speaker, rather than the lexical content of the speaker.

Speech recognition technology is involved in the fields of signal processing【信号处理】, pattern recognition 【模式识别】, probability theory【概率论】, information theory【信息论】, artificial intelligence, and so on.

Driverless car —— the case of integration of perception and control

Driverless Car【无人驾驶汽车】 is a kind of intelligent vehicle, which is also a comprehensive application of image and sound recognition. It mainly relies on a variety of sensors or sensing devices to get the outside information, and then automatically plans route and controls the car (also named wheeled robot【轮式机器人】) behavior. It has been put into use in many countries. Here, we introduce some driverless cars developed by different companies.

1. Google driverless car

Sebastian Thrun, the co-inventor of Google Maps Street View service【Google 街景地图】, presided over the Google Driverless Car project. In 2005, he led a team at Stanford Artificial Intelligence Laboratory【斯坦福人工智能实验室】to design a robotic car. Thrun's robotic car "Stanley" won the Grand Challenge held by the United States Department of Defense Advanced Research Projects Agency (DARPA)【美国国防部高级研究计划署】due to its traversed distance of over 132 miles (212.43 kilometers) in the desert. Google's driverless car is shown in Figure 15.2.

2. ULTra

ULTra (Urban Light Transit)【城市轻轨】 is a PRT

Figure 15.2 Google's self-driving car

(Personal Rapid Transit) pod system, also referred to as PAT (Personal Automated Transport)【个人自动交通】or ATN (Automated Transit Networks)【自动化交通网】developed by the British engineering company ULTra Global PRT with the association of the University of Bristol. The first system began passenger service at London Heathrow Airport in 2010 as a taxi service (Figure 15.3). This car may knock out【淘汰，使出局】buses and make them an outdated transportation. This car looks like a bubble shaped, alien ship, and is powered by batteries. The passengers usually select their destination by using a touch screen【触摸屏】. These electric-powered cars designed to travel up to 40 kilometers per hour (25 miles per hour). Once the destination has been chosen, the control system records the request, and sends a message to the car cabin. Then the car cabin follows an electronic pathway.

Figure 15.3　An ULTra Pod

3. Tubenet transit

Tubenet Transit【小分支（支线）交通，城市轻轨】is being designed and developed in Beijing by the Tubenet Transit Institute led by Dr. Nanzheng Yang (Figure 15.4). This tubecar system operates in a tube using solar energy. The team is awarded a Patent in China as well as in many other countries worldwide. It is an advanced transportation system to achieve zero emission, noiseless and dust free operation, and reliable control and safety system. It is highly efficient, comfortable, fast, economical, and green travel.

Figure 15.4　Tubenet Transit being developed in Beijing, China

4. CyCab

INRIA, a French company, spent ten years of effort to develop a driverless car, named "Kabo race"

(CyCab), which looks like the future of the golf car (Figure 15.5). The equipped Global Positing System (GPS)【全球定位系统】 is similar to that used to cruise missile guidance【巡航导弹制导】. The user interface is a touch screen. In order to know the position of CyCab in real time, it has an improved GPS system, which has an accuracy of a few centimeters. These vehicles can communicate through the Internet, which means the cars can exchange information with each other, and is able to avoid traffic jams and obstacles in their paths.

Figure 15.5 CyCab – a small autonomous car

5. Lux

The car Lux was developed by a German company. It can use several laser scanners to detect the road within 180 meters from different angles. It can construct a 3-D model by GPS road navigation system. In addition, it can also identify traffic signs, and ensure the car's safe driving in the premise to obey the traffic rules.

Smart unmanned aerial vehicle 　— 　the case of integration of perception and control

The Unmanned Aerial Vehicle (UAV)【无人飞行器（无人机）】 is an aircraft with no pilot on board. UAVs can be used as target and decoy【诱饵】 for military maneuver【军事演习】; as surveillance, intelligence, and reconnaissance tool【侦查工具】; in combat operations; for logistics; and as a model in the research of aircraft. Completely autonomous UAVs, also known as smart UAVs, can independently complete a specific task without the help of artificial instruction. It can be used in state monitoring, environment information collection, data analysis and corresponding response. China has been developing diverse range of UAVs in recent years. Some of the Chinese Unmanned Combat Air Vehicles (UCAVs)【中国的无人战斗机器人】 are CASC CH-3 (2012), Guizhou Sparrow Hawk II (2016), NORINCO Sharp Eyes III (2015), NORINCO Sky Saker (2016), and so on.

A quadrotor robot aircraft, designed by Raffaello D'Andrea of TEDGlobal Robotics Laboratory 【TEDGlobal 机器人实验室】, can actually perform automatic tossing and catching ball, and maintain the balance from rotation in the air (Figure 15.6). It can also carry a cup to run around and let the cup stable, and even when the cup is replaced with a vertical stick, it can still maintain the stability, just like a circus. Its perfect combination of sophisticated mathematical model, techniques of mechanical and electrical, intelligent perception and feedback control, is indeed amazing.

Figure 15.6   A quadrotor robot aircraft

(Courtesy of Courtesy of http://www.5imx.com/portal.php?mod=view&aid=490)

The autonomous【自治的】 flight of UAV needs the camera, image perception, algorithm, and complex navigation technology【复杂的导航技术】. By means of 3D camera, the complex environment information is transferred to CPU for fast computation and makes the corresponding judgment. The Real Sense computing【实感计算】 technology of Intel can make the UAV fly over mountains without any GPS or artificial navigation.

15.3.2   Natural language understanding

Image and sound are the natural forms of information, while a language is a meaningful and artificial sequence of information, which may sometimes employ sound or image as its carrier. Enabling a machine to understand natural language will be significant in real life. For example, it can provide users with consulting services【咨询服务】.

The natural language understanding includes several steps: speech recognition【语音识别】, syntactic analysis【句法分析】, semantic analysis【语义分析】, and pragmatic analysis【语用分析】.

In speech recognition, a speech signal is analyzed and is converted to a sequence of words, i.e. the input to the speech recognition subsystem is a continuous (analog) signal and the output is a sequence of words.

In syntactic analysis, the recognition system should know the syntactic structure of natural language, and should be flexible enough to support embedded language at the same time. This is a difficult task because a word may have different parts of speech【词性】 and consequently be used as a different grammatical unit, or it has different meanings【词义】.

After syntactic analysis, the process of semantic analysis can obtain the meaning of a sentence, which creates a representation of the objects involved in the sentence, their relations and their attributes. In the process, any knowledge representation schemes can be used.

Natural language is inherently ambiguous, meaning that the same syntactic structure could have multiple valid interpretations. These ambiguities can arise for several reasons.

Example 1

Time flies like an arrow.

The first example demonstrates lexical ambiguity【单词歧义】. The word "time" can be a noun, or a verb. As a verb it means "arrange", "test", "adjust", "measure", etc. The word "fly" can be a noun, which refers to a small insect with two wings. The presence of two or more possible meanings within a single word is called lexical ambiguity.

Example 2

This is a sewing machine.

This sentence seems to be feasible in our daily life, but may have two feasible meanings: ① This is a machine for sewing; ② This is a machine which is sewing. However, without background knowledge, a computer cannot obtain the real meaning. This case is named syntactic ambiguity【句法歧义】.

Example 3

The dish fell on the bowl and it is broken.

This example is the case of referential ambiguity【指称模糊】. Can we infer that the dish is broken or the bowl is broken or both of them are broken? The sentence is referentially ambiguous because it can refer to both the dish and the bowl. Usually, another step, pragmatic analysis, is required to further clarify the interpretation of the sentence and to remove ambiguities.

Natural language comprehension is a huge area of study and goes well beyond the scope of this book. But it's important to understand the reasons why this issue is so challenging.

15.3.3 Intelligent robot

Intelligent robot technology is essentially an integration of all of the above technologies and the comprehensive use of electronic, mechanical and other technologies. However, in real application, the value and the benefits are also amazing. From the industry viewpoint, we can divide the area into industrial robots, military robots, service robots, and so on.

1. Industrial robots

With rising labor costs, the tendency in modern manufacturing industry to employ automation techniques to boost productivity and to become more efficient and competitive has also increased【随着劳动力价格的上涨，现代制造业利用自动化技术来提高产能，从而变得更有效，更有竞争力的趋势正在增长】. The launch of Industrial Internet by the United States, the concept of "industry 4.0"【工业 4.0】launched by Germany, robots, digital manufacturing, and other technologies set off a new round of Industrial Revolution【发起了新一轮的工业革命】.

Industrial robots with varying degrees of autonomy are developed. Some robots are programmed to carry out specific tasks. Some robots are much more flexible and utilize sensing and conception technology as well as some feedback for the workpiece.

According to the data of IFR【国际机器人联合会】, in 2013, the number of robots in China increased by 36500 units, surpassing other developed countries for the first time in the world. However, the Chinese robot stock is still relatively small, only more than 9 units. However, the world industrial robots have reached more than 1 million, and are mainly distributed in Japan, the United States, Germany and other developed countries. In automotive manufacturing industry, for every ten thousand production workers, the number of robots occupied by Japan is 1710, Italy is 1600, France is 1120, Spain is 950, the United States is 770, the United Kingdom is 610, and China is less than 90 units.

Incorporating industrial robots, some surprising results are achieved. In Shanghai Yaohua

Pilkington Glass corporation【耀皮玻璃（上海耀华皮尔金顿玻璃股份有限公司）】, the automotive production of glass only needs 19 seconds. The new automated factory of Galanz【格兰仕】has achieved "second build" microwave oven (i.e. build a microwave oven within 1 second. In SKU360, new automated logistics distribution center, located in Songjiang Shanghai 【上海，松江（区）】, after the completion of first phase of the project, the daily order processing capacity can reach 1 million 200 thousand pieces.

2. Military robots

Military robots【军事机器人】 are designed for military applications, from transport to search & rescue and attack to protection of soldiers and people in the field.

Cheetah robot【"猎豹"机器人】, developed by Boston Dynamics【波士顿动力（公司）】in 2012, is capable to sprint, take sharp turns, and stop suddenly. It runs at speeds of 29 mph on a treadmill.

WildCat【"野猫"】, a free-running version of Cheetah developed in 2013, is capable of running outdoors untethered【不受限制的】at speeds of 16 mph on flat terrain. Both Cheetah and WildCat were developed for United States military (Figure 15.7).

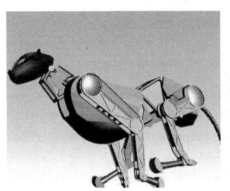 

Figure 15.7   (a) The Cheetah robot, (b) The WildCat robot

(Courtesy of http://www.bostondynamics.com)

Anti-tank robot is equipped with anti-tank missiles, TV camera, and laser ranging machine【激光测距机】, and controlled by a microcomputer and a control system. Its shape is similar to a minivan【小型货车】with remote control. On finding the target, the robot can self-maneuver【自（演习）执行】or is operated by distant remote control. It can occupy a favorable firing position, determine the firing elements by laser ranging, and then aim at target missile.

Tactical reconnaissance robot【战术侦察】 is a humanoid intelligent robot【具有人的特点的智能机器人】. Its small body is equipped with infantry reconnaissance radar【步兵侦察雷达】or infrared, electromagnetic, optical, and audio sensors and radio and optical fiber communication equipment【红外、电磁、光学、音频传感器和无线电和光纤通信设备】. It can self-operate and perform independent observation and reconnaissance. It can also been airdropped, or thrown into the depth of the enemy to reach a proper position for setting up a Reconnaissance.

The robot BigDog【"大狗"机器人】 is also developed by Boston Dynamics Corporation【波士顿动力公司】(Figure 15.8). The quadruped robot【四足机器人】can quickly restore the balance after an impact and continue on its route. It can be used as a mule to perform tasks in mountains or in various terrains filled with obstacles. This machine dog can play a very important role on the battlefield: to carry ammunition【弹药，军火】, food and other items for soldiers in the area of traffic inconvenience. It can not only walk and run, but can also cross a barrier of certain height. The power of the robot comes from a hydraulic【水压的】 system of gasoline【汽油】 engine.

Figure 15.8 The BigDog robot

(Courtesy of http://www.bostondynamics.com)

3. Service robots

In recent years, the global market for service robots maintains a rapid growth rate. According to data from the International Federation of Robotics【国际机器人联合会】, in 2011, the sales of service robots for professional use were about 16,400, 9% more than in 2010, and the sales of service robots for personal and domestic use were 1.7 million.

Under the support of national 863 plan【国家 863 计划】, China has carried out a lot of work, and achieved certain results in the area of service robot research and product development, such as Harbin Industry University【哈尔滨工业大学】 developed the tour guide robot【导游机器人】 and cleaning robot【清洁机器人】; South China University of technology【华南理工大学】developed the nursing robot【护理机器人】; Institute of automation and Chinese Academy of Sciences【中国科学院自动化研究所】 developed intelligent wheelchairs【智能轮椅】,etc.

15.4 Reasoning

15.4.1 Searching

Solving a problem can be taken as searching a proper solution or several solutions in a state space. A proper state is corresponding to a solution. In Chess【国际象棋】 or Go【围棋】, each position can be taken as a state. Go is a board game like Chess, which originated in China, and is popular in a number of countries. An intelligent computer can test each position and evaluate it, and then choose the most optimal next position. This progress is essentially searching in the state space. Other general problems are similarly converted to search problems.

Search Methods

There are two general search methods: brute-force search【暴力(强力)搜索】and heuristic search【启发式搜索】.

Brute-force search, also called blind search and uninformed search, can be described in one sentence: "Try every step (state) till the results are found". For example, give a set of numbers and find all prime numbers.

Breadth-first search【广度优先搜索】and depth-first search【深度优先搜索】are the brute-force techniques in which a special data structure is utilized. The data structure can be a tree, a graph, or other, which has been discussed in the previous chapter.

Heuristic search【启发式搜索】, also called informed search, uses some outer hints or directions. For example, the solution may be known in a subset of search space, and the algorithm performs searching only within the subset.

It is well known that Deep Blue【"深蓝"(机器人)】, a chess-playing computer【国际象棋玩机】developed by IBM, defeated world class champion, Garry Kasparov【加里·卡斯帕罗夫】, on February 10, 1996 under regular time controls. Deep Blue was then heavily upgraded and played Kasparov again in May 1997, and won the six-game rematch of 3½–2½, under standard chess tournament【象棋锦标赛】time controls. This proved to be a landmark【标志性】event in the history of artificial intelligence. Deep Blue versus Garry Kasparov match is still cited as the symbolic turning point. The algorithms of Deep Blue were based on brute-force search【暴力搜索】. It can prejudge each possible chess position, evaluate it, and then choose the optimal position. Within one second, it can evaluate 200 million chess positions. The architecture used in Deep Blue was applied to financial modeling, data mining, and molecular dynamics【分子动力学】. It is a massively parallel powerful supercomputer【大规模并行超级计算机】 with 30 nodes, with each node containing a 10 MHz P2SC microprocessor, enhanced with 480 special VLSI (Very Large Scale Integration)【超大规模集成电路】 chess chips【480个专门的超大规模集成电路象棋芯片】. However, the focus of chess play has now been shifted to software chess programs【转向象棋软件程序】 rather than using dedicated chess hardware【专门的象棋硬件】.

In 2006, chess program【象棋程序(软件)】Deep Fritz won 4-2 world chess champion Vladimir Kramnik【弗拉基米尔·克拉姆尼克】. The algorithm of Deep Fritz was based on heuristic search【启发(式)搜索】.

In February 2011, a Robot machine named IBM Watson【沃森】 beat the champions at a more complicated game, Jeopardy【绝境】.Watson demonstrated that a whole new generation of human-machine interactions【人机交互】 will be possible. The algorithm used for Jeopardy is QA (Question Answering【问答】), which is part of algorithm IR (Information Retrieval【信息检索】).

About twenty years later after the victory of Deep Blue, in October 2015, AlphaGo【阿尔法狗;阿尔法围棋机】, another Go chess-playing computer【国际围棋玩机】 developed by David Silver, Aja Huang, and their research group, a Google DeepMind team【Google "深思" 团队】, won the same honor. In March 2016, it defeated Lee Sedol【李世石】, a famous South Korean professional Go player of 9-dan rank【韩国职业九段围棋高手】, without handicap; a feat that was previously believed to be at least a decade away. The AlphaGo used many intelligent techniques including Monte Carlo【蒙特卡洛】 tree search, neural networks【神经网络】, and reinforcement learning【增强式学习】. AlphaGo's algorithm use heuristic search method【启发式搜索方法】.

15.4.2 Logic

1. Propositional logic

Propositional logic【命题逻辑】 is a logical language which is similar to our natural language. In propositional logic, symbols are used to represent logical representation and operators are modifiers of symbols, and are used to connect them, just like conjunctions or adverbs in our natural language.

Let A and B be the propositions (statements in a natural language), then a sentence in propositional language is defined as:

A ↔ B   (logical equivalence【逻辑等价】), means "A if and only if B"
A → B   (implication【蕴含】), means "if A then B"
A ∨ B   (disjunction【(逻辑) 或,析取】), means "A or B"
A ∧ B   (conjunction【(逻辑) 与,合取】), means "A and B"
~A   (negation【(逻辑) 非】), means "not A"

According to the meanings listed above, we can specify the truth table for each logical operator in propositional logic, as shown in Table 15.1.

Table 15.1 Truth table for logical operators in propositional logic

A	B	A ∧ B	A ∨ B	A → B	A ↔ B	~A
F	F	F	F	T	T	T
F	T	F	T	T	F	T
T	F	F	T	F	F	F
T	T	T	T	T	T	F

How to understand the values in the table? We can start from an example:

A: "I have some cash to buy a ticket."

B: "I will go to see the film."

A→B: "If I have some cash to buy a ticket, then I will go to see the film."

We can take the expression A→B as an assertion【断言】: "If A then B" is false if the assertion is broken, otherwise it is true. In Table 15.1, the last two lines can be easily understood. The last line means: "If I **truly** have some cash to buy a ticket, then I will **truly** go to see the film", which can verify the logical expression: A→B. The third line means "If I **truly** have some cash to buy a ticket, (but) I will not go to see the film **in fact**", which is contradictory to this assertion, so the value of the expression is "F" (false). The first line is also not difficult. If I **truly have no cash** to buy a ticket, (so) I will not go to see the film, which is reasonable of course.

But for the second line, how to understand? "I **truly** have no cash to buy a ticket, but I **truly** go to see the film." How? May be I have a credit card, or use AliPay directly? Does this fact break the rule: A→B? No, so F→T is "T" (true).

Okay, now, the most difficult truth table has been achieved, the others are relative easy. Can we construct them? (See exercise.)

A sentence in propositional logic is recursively defined as shown below:

(1) An uppercase letter, such as "A", or "B", is used for a statement in a natural language, referred to as a sentence in propositional logic.

(2) Any of the two constant values (true and false) is a sentence.

(3) If A is a sentence, then ~A is a sentence.

(4) If A and B are sentences, then $A \wedge B$, $A \vee B$, $A \rightarrow B$, and $A \leftrightarrow B$ are sentences.

Examples

The following are sentences in propositional logic:

(1) Today is Friday (F).

(2) It is sunny (S).

(3) Beijing is the capital of China and Guangzhou is the capital city of Guangdong province ($B \wedge G$).

(4) Today is not holiday (~H).

(5) If the weather is pleasant, then I will play outside ($P \rightarrow O$).

In propositional logic, the process of deduction is the creation of new facts and logical expressions from the existing facts. For example:

The weather is fine today.	Premise 1:
If the weather is fine, then we'll go to play football.	Premise 2:
Therefore, we'll go to play football.	Conclusion

If we use A for "*The weather is fine today*", B for "*we'll go to play football*", and the symbol |– for "therefore", then we can show the above argument as:

$$\{A, A \rightarrow B\} \vdash B$$

2. Predicate logic

Predicate logic【预测逻辑】 defines the relation between the parts in a proposition, so we can summarize: propositional logic is about sentence level, and predicate logic is about word level.

In predicate logic, a sentence is divided into a predicate and some parameters. The parameters, just like the subject and object of a sentence, are in natural language. For example,

P1: "*Kathy is Angela's mother*"	=	Mother(*Kathy, Angela*)
P2: "*Anais is Kathy's mother*"	=	mother(*Anais, Kathy*)

The predicate "mother" represents the relationship of motherhood in each of the above sentences. We can deduce a new relation between Anais and Angela provided the object Kathy in each of the sentences refers to the same person: *grandmother* (Anais, Angela).

A predicate becomes a proposition when we quantify it. Two quantifiers are common in predicate logic: universal quantifier and existential quantifier. The universal quantifier, $\forall$, read as "for all", represents that something is true for all objects that its variable represents. The existential quantifier, $\exists$, read as "there exists", represents that something is true for one or more objects that its variables represents.

Examples

(1) Ancient Greek philosopher Aristotle gave an example of logical reasoning, which can be written in predicate logic as shown here:

"All men are mortals" can be written as: $\forall x[man(x) \rightarrow mortal(x)]$

（2）The sentence "Some swans are white" can be written as: $\exists x[swan(x) \wedge white(x)]$

（3）An example of inference:

All men are mortals　　　　　　　　　$\forall x[man(x) \rightarrow mortal(x)]$

Socrates is a man　　　　　　　　　　$man(Socrates)$

Therefore, Socrates is mortal　　　　　Conclusion

We can replace one instance of the class man (Socrates) in the first sentence to get the following argument:

$$\forall x[man(x) \rightarrow mortal(x)], man(Socrates) \vdash mortal(Socrates)$$

3. Fuzzy logic

Fuzzy set theory【模糊集理论】is a means of specifying how well an object satisfies a vague description. For example, consider the proposition "The temperature is very high today." Is this true if the temperature lies between 30°C and 35°C, or even 40°C? It is hard to distinguish "true" or "false". Fuzzy set theory treats it as a fuzzy predicate and defines a function with a value between 0 and 1, rather than as just true or false, such as for the temperature problem, the function is $Height(t)$, where the parameter t is the temperature value. Fuzzy logic is based on fuzzy set theory, and defines a fuzzy function for each proposition to represent the degree of truth. Consequently, the logic expression can be defined as:

$$T(P \wedge Q) = \min(T(P), T(Q))$$
$$T(P \vee Q) = \max(T(P), T(Q))$$
$$T(\sim P) = 1 - T(P)$$

There are some other non-classic logistics【非经典逻辑】, such as modal logic【模态逻辑】, multiple-valued logic【多值逻辑】, and so on.

15.4.3　Data mining and machine learning

Data mining【数据挖掘】is the finding of feasible information from related data and then applying mathematical analysis to derive valuable patterns and trends. Usually, Data mining system achieves the objectives by statistics, online analytical processing【在线分析处理】, information retrieval【信息抽取】, machine learning【机器学习】, expert systems (that rely on past experiences), pattern recognition【模式识别】, and many other methods.

Machine learning mainly studies how the computer simulates the learning behavior of human beings. In general, machine learning systems can improve their performance by acquiring new knowledge or skills, and re-organizing the existing knowledge structure. The field is the core of artificial intelligence, and is the basic way of making computer to become more intelligent. Application of machine learning has spread in all branches of artificial intelligence, such as expert systems, automated reasoning, natural language understanding, pattern recognition, computer vision, robotics and other fields.

15.5　Nature Inspired Computation

15.5.1　Introduction

Natural Computation (Nature Inspired Computation)【自然计算】can imitate nature, and usually build models or algorithms with adaptive, self-organizing【自组织】, and self-learning【自学习】ability to solve the problems which are difficult to be solved by traditional calculation methods. The application field of natural computation includes complex optimization problems, intelligent control, pattern recognition, neural networks, network security, hardware design, quantum computing, evolutionary algorithms, and the ecological environment of the application【自然计算的应用领域包括复杂的优化问题、智能控制、模式识别、神经网络、网络安全、硬件设计、量子计算、演化算法和生态学】. Table 15.2 shows summary of some models or algorithms of Natural Computations.

Table 15.2 Summary of some models or algorithms of Natural Computations

Inspired from	Models or Algorithms
Brain information processing	Artificial Neural Network【人工神经网络】
Fuzzy way of thinking	Fuzzy System【模糊系统】
Biological immune mechanism	Artificial Immune【人工免疫】System
Biological evolutionary process	Evolutionary Computation (EC)【演化计算】

15.5.2 Artificial neural networks

An artificial neuron【人工神经元】is just like a single biological neuron【生物神经元】. It sums some weighted inputs, and then compares the result with a threshold value. If the result is larger than the threshold value, the perceptron fires, otherwise it does not fire. So it is named as perceptron【感知器】, which is the simplest neuron network. When the perceptron fires, the output is 1, and when it does not fire, the output is zero.

How to use it? For example, we can imagine that if forest fires may be generated by multiple factors, such as temperature, humidity, wind, and so on. How the sensors alarm? The multiple factors may be represented by $X_1, X_2, \ldots, X_n$, and the output Y can be the alarm information (Figure 15.9).

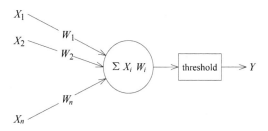

Figure 15.9   The input and output of a perceptron

Early neural networks, called feedforward neural networks【前馈神经网络】, have no feedback mechanism. However, the weight values $W_1, W_2, \ldots, W_n$ can be modified dynamically to adapt to the changing situation. Machine learning technology can evaluate the output value Y and the input values $X_1, X_2, \ldots, X_n$ to improve the value of the weight values $W_1, W_2, \ldots, W_n$. This type neural network is called recurrent neural network (RNN)【反馈神经网络】.

Multi-layer networks

Let's imagine what the result would be if the Y port of the neuron was connected to the X_i of another neuron. The second neuron can also been connected to the third one. Obviously, we can use this mechanism to construct multi-layer neural networks【多层神经网络】, as illustrated in Figure 15.10. The output of each layer becomes the input of the next layer. The input layer is the first layer, the hidden layers are the middle layers, and the output layer is the last layer.

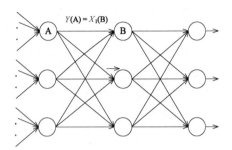

In real applications, many kinds of neural networks work together to accomplish the related tasks. Such as the intelligent system Alpha Go, that uses two different neural networks. The policy network responsible for assessment of next position chooses to reduce the search scope and value

Figure 15.10 Schematic diagram of multi-layer neural network

network is responsible for the assessment of our odds【胜算】 to reduce the depth of the search. The schematic diagram of multi-layer neural network is shown in Figure 15.11.

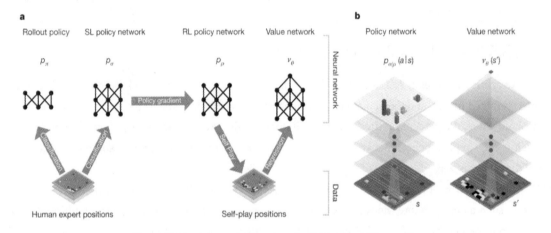

Figure 15.11 Schematic diagram of multi-layer neural network

(Courtesy of D Silver et al. Nature 529, 484-489 (2016) doi: 10.1038/nature16961)

15.5.3 Bionic intelligence

The neural network is just the basis of bionic【仿生学的】 calculation. Bionic intelligence【仿生智能】 can simulate biological activities in the areas including artificial immune system【人工免疫系统】, artificial endocrine system【人工内分泌系统】, artificial life【人工生命】, and so on. However, simulation to human brain is the most worth mentioning.

Known as one of the world's six scientific experiments【作为世界上六大科学实验之一】, the Blue Brain Project【蓝脑计划（工程）】 is a masterpiece of bionic intelligence. The project is headed by a Swiss researcher, Henry Markram. Its basic function is to use computer to simulate activity of the human brain, in order to study its various functions such as cognition, feeling, memory, thinking, etc. However, the primary purpose of the project is for the treatment of Alzheimer's disease【爱默生病（老年痴呆症）】 and Parkinson's disease【帕金森病（震颤性麻痹）】. By accurately simulating the activity of a brain, scientists can also uncover the secret hidden in the insane. However, the generated results may be far more than only used for the treatment of diseases.

Blue Brain project simulates neocortical columns (NCC)【新大脑皮层】by software. The software model of the project is based on the experimental data for 15 years. For the analysis and processing of the information and the automatic reconstruction of physiology correct neurons, the corresponding 3-D model has been established and many related software tools have been developed.

The blue brain project began in 2005 and its aim was to study the unique structure of the mammalian【哺乳动物】 brain. It initially focused on simulation of the NCC of rats containing 10000 neurons. In humans, each NCC contains 60000 neurons. In July 2009, the project finished the calculation of human NCC and draw out a simulation of neuronal activity in the 3D map. The project is currently continue, and expected by around 2020 to make first "thinking" machine in the scientific history, which will possibly be have the feeling of desire, pain, fear, and other emotions.

15.5.4 Evolutionary computation

Evolutionary computation (EC)【演化计算】 follows the biological evolution in nature, and based on the principle of survival of the fittest. It represents a variety of complex structures with simple code, and then simulates genetic operations by simple code operations【它用简单的编码表达大量的复杂结构，并且也通过简单的编码操作来模拟基因的操作】. It simulates the survival of the fittest competition mechanism by searching the solution space【它通过搜索解空间来模拟适者生存的机制】. EC has the characteristics of self-organization, self-adaptation【自适应】, self-learning【自学习】, and intrinsic parallelism【本质上（内在）的并行】. It is not restricted by the restriction of the search space or any other auxiliary information【它没有搜索空间的限制，也无需辅助信息】. Therefore, evolutionary algorithms are simple, general, easy to operate, and highly efficient. EC can solve large scale optimization problems, such as machine learning, adaptive control【（自）适应控制】, artificial life【人工生命】, artificial neural network, and economic forecasting【经济预测】. It has achieved success in various application fields, including mathematics, physics, chemistry, biology, computer science, social science, economics, and engineering.

Genetic algorithms (GAs)【遗传（基因）算法】 are used to conduct search in a problem space. The basic idea is to express the problem as a search space, and the potential results are modeled as genes【基因】, which can be realized as variables with some data structure in a program. Evolutionary algorithms (EA) can produce different genes by simulating gene selection, crossover, inheritance, and mutation. This process is continued until a termination condition is reached to realize optimal search.

Now, the research on the calculation of the content is very extensive and the design and analysis of EC is the theoretical basis for the calculation and application in various fields. With the continuous expansion of the research on the theory of EC and the deepening of the application field, evolutionary computing will achieve great success and occupy more important position in today's society.

In recent years, there have been some new evolutionary branches, such as non-Darwin theory of evolution 【非达尔文的演化理论】, and memetic algorithm【模因（或文化基因）算法】 (For more details, refer to Koza, J). Summary of some ECs is shown in Table 15.3.

Table 15.3 Summary of some evolutionary computations

Algorithm	First Introduced by	Main Ideas or Objective
Genetic Algorithms (GA)【基因算法（或遗传算法）】	John Holland, in 1970s	To use three operators (selection【选择】, crossover【交叉】, and mutation【变异】) to process some data structures which are used for simulating biological gene to get the result of the problem. The largest application of techniques is in the domain of optimization, which are the mainstream algorithms of EC
Evolution Strategies (ES)【演化策略】	I. Recenberg, H. P. Schwefel, in 1960s	To solve parameter optimization problems
Evolutionary Programming (EP)【演化规划】	L. J. Fogel, in 1960s	To evolve Finite State Machines (FSM)【有限状态机】 to predict events on the basis of former observations
Genetic Programming (GP)【基因（遗传）规划】	Cramer, in 1985. J, Koza, make it more perfect, in 1992	To evolve the program itself. EA + PS = AP (Automatic Programming)

15.5.5 Swarm Intelligence

Ant colony optimization (ACO)

Ants in the process of finding an unknown food can be demonstrated as follows. When an ant finds some food, it releases a volatile secretion【挥发性分泌物】, named pheromone【信息素】, which gradually evaporates【蒸发掉】 with the passage of time. The pheromone concentration size reflects the length of the path. Because of the attraction of pheromone, more and more ants will find food, and release more information, resulting in strengthening of the information. Some ants do not traverse the same path and follow a new path. If the new path is shorter than the original path, then more ants are attracted to the shorter path. Eventually, after a period of time, it may appear that the shortest path is the one which is repeated by most of the ants.

Ant colony algorithm【蚁群算法】 is an inherently parallel algorithm. The process of an ant search is independent of each other, and only the pheromone is used for communication. So the ant colony algorithm can be viewed as a distributed multi-agent system, as it begins to search in different positions simultaneously. This process is independent of the solution search. This mechanism not only increases the reliability of the algorithm, but also makes the algorithm a strong global search algorithm.

Particle Swarm Optimization (PSO)

Particle swarm optimization (PSO)【粒群优化】 algorithm is also an EC algorithm, which originates from the behavior of flocks of birds. The algorithm is inspired by the activities of cluster of birds. By sharing information and making the whole group as a dynamic evolution from disorder to order in the problem space, the algorithm can obtain the optimal solution.

PSO is similar to genetic algorithm, and it is also a kind of optimization algorithm based on iteration. Initially, the system creates a set of random solutions, and then searches for the optimal value by iteration and adjusts each particle according to the most optimal particle. However, it has no crossover and mutation operators of genetic algorithm. Compared with the genetic algorithm, the advantages of PSO are simple and easy to be realized and the algorithm has no need to adjust many parameters. Currently, it has been widely applied in function optimization, neural network training, and fuzzy system control applications.

15.6 References and Recommended Readings

- Chen M, Mao S, Zhang Y, Leung V C. Big Data: Related Technologies, Challenges and Future Prospects. Springer Briefs in Computer Science, Springer, 2014.
- Luger G F. Artificial Intelligence: Structures and Strategies for Complex Problem Solving. MA: Addison-Wesley, 2008.
- Stuart R, Peter N. Artificial Intelligence: A Modern Approach. Upper Saddle River, NJ: Prentice Hall, 2003.
- Nilsson N J. Principles of Artificial Intelligence. Tioga Publishing Company, CA: Palo Alto, 1980.
- Ross T J. Fuzzy Logic with Engineering Applications. John Wiley & Sons, 2010.
- Mitchell T M. Machine learning. Beijing : China Machine Press , 2003 (in Chinese).
- Goldberg D E. Genetic Algorithms in Search, Optimization & Machine Learning. Boston, MA: Addison-Wesley, 1989.
- Holland J H. Adaptation in Natural and Artificial Systems. Massachusetts Institute of Technology, MIT Press, 1992.
- Michalewicz Z. Genetic Algorithms + Data Structures = Evolution Programs. Springer-Verlag, 1996.
- Schwefel H P. Evolution Optimum Seeking, Sixth-Generation Computer Technology Series, Wiley, 1995.
- Fogel L J. Autonomous Automata. Industrial Research, vol.4, pp.14-19, 1962.
- Koza J R. Genetic Programming. On the Programming of Computers by Means of Natural Selection, MA: MIT Press, 1992.
- http://en.wikipedia.org/wiki/Memetic_algorithm.
- Bonabeau E, Dorigo M, Theraulaz G. Swarm Intelligence: From Natural to Artificial Systems. New York: Oxford University Press , 1999.
- Wu W T. Basic Principles of Mechanical Theorem Proving in Elementary Geometries. Journal of Automated Reasoning, vol. 2(3), pp. 221-252, 1986.
- https://en.wikipedia.org/wiki/Wu's_method_of_characteristic_set.
- Wen-Tsun Wu Work in geometric theorem proving: Wu's method, Herbrand Award 1997.
- Russell S J, Norvig P. Artificial Intelligence: A Modern Approach. Upper Saddle River, New Jersey: Prentice Hall, 2010.
- Franklin S, Graesser A. Is it an Agent, or Just a Program?: A Taxonomy for Autonomous Agents. Proceedings of the Third International Workshop on Agent Theories, Architectures, and Languages, Lecture Notes in Computer Science, Springer, vol. 1193, pp. 21-35, 1997.
- Silver D, Huang A, Maddison C, etc. Mastering the game of Go with deep neural networks and

tree search. Natrue, vol. 529(7587), pp. 484-489 (Published online 27 January 2016).
- Wikipedia, https://en.wikipedia.org/wiki/Deep_Blue_(chess_computer), retrieved on March 18, 2016.
- TedGlobal corp: http://www.ted.com, retrieved on March 27, 2016.
- Boston Dynamics corp: http://www.bostondynamics.com, retrieved on March 27, 2016.
- The Blue brain project: http://bluebrain.epfl.ch/, retrieved on March 28, 2016.
- Markram H: The Blue Brain Project, Nature Reviews Neuroscience, vol. 7, pp.153-160, 2006.
- http://www.chinabaike.com/article/sort0525/sort0541/2007/20070730156620.html.
- http://www.ifr.org/【国际机器人联合会】.
- http://tech.sina.com.cn/other/2003-08-26/1159225499.shtml.
- http://www.zhongtian-auto.com/en/news-zhongxin/yenei-news/china-gongyejiqirenchanyehuafazhanzhanlue.html.

15.7 Summary

- Artificial intelligence is the study of programmed systems that can simulate, to some extent, human activities such as intelligently behaving, perceiving, reasoning, decision-making, problem-solving, thinking, learning, and acting.
- There are three schools in the history of artificial intelligence. The symbolism school deems the human cognition primitive is symbols, and cognitive processes are symbol operation. The Connectionism school thinks of neurons as primitive. The behaviorism school thinks intelligence depends on perception and action.
- The first step in applying artificial intelligence is knowledge representation, which includes semantic network, rule system, formal logic and other expressions. It will be used in perception, reasoning and some intelligent algorithms which are the kernel technology in this area.
- Intelligent Agent is an abstract concept. It generally refers to the object with intelligent techniques. At the same time, it is also the subject of executing the intelligent algorithm.
- Perception can make the agent hear and see the world, reasoning can make the agent think, and the intelligent computing can make the agent become smarter.
- Data mining and machine learning make the system have the ability to go beyond the basic logical reasoning, which can be taken as complex intelligence.
- In recent years, a great progress has been made in the simulation of biological and nature's phenomena. This simulation can be either a structure, such as neural network, or a thing of the intrinsic transformation process, such as genetic algorithm, or the nature of many objects of motion, such as particle swarm algorithm.

15.8 Practice Set

1. Using the symbols Revolves_around, Rotates, Is, In, Planet for the predicates "revolve around",

"Rotates on its axis", "is", "in", "is a planet" respectively, write the following sentences in predicate logic:
- a. The earth revolves around the sun.
- b. Venus rotates on its own axes.
- c. Venus is a planet.
- d. There are some planets in the solar system.
- e. The earth does not revolve around Jupiter.
- f. Not every planet is solid.
- g. The earth revolves around the sun, and also rotates on its axis.
- h. A celestial body is a planet if and only if it revolves around a star.

2. A perceptron: _____.
   - a. is an artificial neuron.
   - b. applies a weight on signals that pass through the neighboring neuron.
   - c. is a biological neuron.
   - d. is one of the parts of a biological neuron.

3. Which of the followings is not a sentence in propositional logic? _____.
   - a. 2.
   - b. If it will rain tomorrow.
   - c. False.
   - d. Where is John?

4. Use a truth table to find whether the following arguments are valid: _____.
   - a. $\{P \rightarrow Q, P\} \vdash Q$
   - b. $\{P \rightarrow Q, Q \rightarrow R\} \vdash (P \rightarrow R)$
   - c. $(P \rightarrow Q) \leftrightarrow (\sim P \vee Q)$
   - d. $\{P \wedge Q, P\} \vdash Q$

5. According to the description below, draw a semantic network. You can use the edges named "is part of" and "is located in".

"Guangdong is a part of China, and Guangzhou is a part of Guangdong. There are four famous mountains in Guangdong province: Xiqiao Mountain, Luofu Mountain, Danxia Mountain, and Dinghu Mountain."

6. Construct the truth table for logical operators in proposition logic.

7. Draw a neural network that can simulate an OR gate.

8. There are three schools of artificial intelligence, which school mainly focus on EC algorithms?

9. Use a truth table to find whether the following argument is valid:

$\{A \vee B, A\} \vdash B$

10. List the steps in natural language understanding.

Unit 16

Internet of Things, Cloud Computing and Data Science

The latest ICT revolution, Internet of Things (IoT)【物联网】, cloud computing【云计算】, and data science【数据科学】 have transformed the way we live by creating new technological breakthroughs. This transformation from digital to real-time intelligence has made IoT, cloud, and data science indispensable to the very nature of work. These technologies have revolutionized our globe and are considered as complementary as they have enhanced each other capacities and capabilities.

16.1 Introduction

16.1.1 Internet of Things

IoT, sometimes called Internet of Everything (IoE), represents a paradigm in which any physical thing can become a computer that is connected to a dynamic and self-configuring global network. IoT can focus on the virtual identity of the smart objects and their capabilities to interact intelligently between different kinds of objects and networks based on interaction and communication interfaces such as Radio-Frequency Identification (RFID)【无线射频识别】 tags, sensors【传感器】, actuators, 5th generation (5G)-enabled smart phones, barcodes【条形码】 or two-dimensional (2D) codes【二维码】. The IoT ecosystem is an extension of the Internet to the physical world through embedded technology and Internet Protocol (IP)【互联网协议】 to create a smart world where objects are integrated into a service-oriented architecture【面向服务体系架构】 of the future Internet【物联网生态系统是互联网到物理世界的延伸，通过嵌入式技术和互联网协议来创造一个万物被集成到一个面向服务体系架构的智能世界】.

Things included in IoT are much more than everyday objects, vehicles, electronic equipment, smart phones, mobile devices, utility meters, cameras, Bluetooth, Wireless Fidelity (Wi-Fi), microwave dish, Worldwide Interoperability for Microwave Access (WiMax), Long Term Evolution (LTE), Satellite, Ethernet【以太网】, Global Positioning Systems (GPS)【全球定位系统】, sensors, RFID, wireless sensor networks (WSNs)【无线传感网络】, Machine-to-Machine (M2M) communication【机器与机器间通

信】, 2G/3G/4G/5G, IP, industrial systems, agricultural systems etc.【物联网所包含的物体远远多于日常生活中的物体，包括车辆、电子设备、智能手机、移动设备、智能电表、照相机、蓝牙、Wi-Fi、微波天线、全球微波互联接入、LTE、卫星、以太网、全球定位系统、传感器、无线射频识别、无线传感网络、机器与机器间通信、2G/3G/4G/5G、互联网协议、工业系统、农业系统等】. Cyber-Physical Systems (CPS)【信息物理系统】 are composed from the integration of embedded computing devices, networks, smart objects, people, communication infrastructure, and physical systems【信息物理系统是嵌入式计算设备、网络、智能物体、人、通信设施以及物理系统的综合】. IoT-based CPS include systems and devices such as robots, implantable medical devices, intelligent buildings, smart cities, smart grids, smart buildings, smart homes, smart factories, smart transport, smart energy, smart industry, smart health, vehicles that drive themselves and planes that automatically fly in a controlled airspace【基于物联网的信息物理系统包括系统和设备，比如机器人、植入式医疗器械、智能大厦、智慧城市、智能电网、智能楼宇、智能家居、智慧工厂、智能交通、智慧能源、智慧产业、智能健康、自动驾驶车辆、在管制空域自动飞行的飞机】 (Figure 16.1).

Figure 16.1　Pilotless plane

　　The technologies that make intelligent IoT applications possible are considered as the enabling technologies【使能技术】 of IoT. The key enabling technologies for IoT are Internet Protocol version 6 (IPv6)【因特网协议第 6 版】, IPv6 over Low power Wireless Personal Area Networks (6LoWPAN)【基于 IPv6 的低速无线个域网标准】, Future Internet 【下一代互联网】, cloud computing, big data analytics【大数据分析】, CPS, semantic technologies【语义技术】, nanoelectronics, system integration【系统集成】, embedded systems【嵌入式系统】, sensor networks, implantable medical devices, coil-on-chip technology, monolithic chip, ZigBee【无线个域网/紫蜂协议】 (used in the context of WSNs), Software Defined Radio (SDR)【软件定义的无线电】, Long Term Evolution (LTE) and LTE-A, cognitive networks【感知网络】, context-aware computing, complex event processing (CEP)【复杂事件处理】, predictive, streaming, and behavioral analytics (Figure 16.2). IPv6 is 128-bit Internet address scheme and its deployment across the world can potentially be used to address any communicating smart object, eliminating the need of network address translation (NAT)【IPv6 是 128 位的互联网地址，可以部署到全世界，能用来标识任何可以进行通信的智能设备，消除网络地址转换（NAT）的需要】. 6LoWPAN is a communication standard that works on the IPv6 protocol suite based on IEEE

802.15.4 standard【基于 IPv6 的低速无线个域网标准工作在 IPv6 协议栈，基于 IEEE802.15.4 标准】. Hence, it allows the low power IoT devices to communicate in a more efficient way【因此，它允许低功耗的物联网设备以一种更高效的方式进行通信】.

IPv6	Future Internet	cloud computing	big data analytics	embedded systems	sensor networks	Bluetooth
Wi-Fi	sensors	RFID	Satellite	Ethernet	GPS	M2M
2G/3G/4G/5G	6LoWPAN	CPS	semantic technologies	nano electronic	system integration	coil-on-chip technology
monolithic chip	ZigBee	SDR	LTE & LTE-A	cognitive networks	context-aware computing	CEP
predictive analytics	smart cities	smart grids	smart buildings	smart industry	smart transport	smart energy
smart health	smart homes	smart vehicles	wearables	actuator	computer vision	in-memory processing
		machine learning	real-time processing	speech recognition		

Figure 16.2　Enabling technologies of IoT

16.1.2　Cloud computing

Cloud computing is a model in which Information Technology (IT)【信息技术】 resources and services are abstracted from the underlying infrastructure and accessed on demand by the customers based on service-level agreements (SLAs)【服务等级协议】 established through negotiations between the cloud service providers (CSPs)【云服务提供商】 and consumers【云计算是将底层的硬件设备抽象成信息技术资源和服务的模型，客户基于服务等级协议（SLAs）按需进行访问，而该协议则是由云服务提供商和客户之间相互协商制定的】. The customers can create or specify their needs according to their requirements in terms of resources, performance, availability【可用性】, etc., and pay only for what they consume【用户可以根据他们对于资源、性能和可用性等的要求创建或指定他们的需求，按使用量付费】. Cloud computing paradigm is a continuous evolution of a client-server technology, virtualization, grid computing【网格计算】, service-oriented applications, and broadband networks【带宽网络】. Cloud computing offers flexible and fair pricing, virtualized network resources, ubiquitous (i.e. reliable/efficient/secure) and optimal resource utilization, scalability【可扩展性】, compatibility【兼容性】, elasticity【弹性】, on-demand resource delivery【资源按需交付】, resource sharing【资源共享】, lower costs, ease of use, environmental sustainability, and management automation【云计算提供弹性和公平的定价、虚拟化网络资源、可靠/有效/安全、优化的资源利用、可扩展性、兼容性、弹性计算、资源按需交付、资源共享、较低的成本、易使用、环境可持续性以及自动化管理】. The cloud computing model is composed of a number of essential characteristics, three service models, and four deployment models, as shown in Figure 16.3. The cloud service can be hosted on-site or off-site such as Tencent cloud【腾讯云】, Alibaba's Aliyun【阿里云】, Xiaomi's Mi Cloud【小米云】, Apple's iCloud, Microsoft's SkyDrive and Samsung's S-Cloud.

Unit 16　Internet of Things, Cloud Computing and Data Science

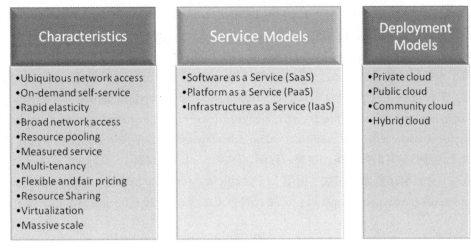

Figure 16.3　Cloud computing model

Customers can lease cloud resources in real time according to their requirements and peak computation times. The resources are usually delivered over the Internet, and run on different virtual machines (VMs)【虚拟机】 side by side on the same physical infrastructure【资源一般通过互联网交付，一起运行在同一物理基础设施上面的不同的虚拟机上】. Thus virtualization provides separation, protection, and efficient resource sharing 【虚拟化技术提供了分离，保护和高效的资源共享】. Science and engineering, data mining【数据挖掘】, computational financing, mathematical applications, gaming, social networking, educational as well as many other computational and data-intensive applications could greatly benefit from cloud computing. Some major cloud providers are Amazon Web Services (AWS), Microsoft Windows Azure, Rackspace, and Baidu Cloud.

16.1.3　Data science

Data science is the study and practice of extracting knowledge from data and finding valuable insights in data to help make valuable decisions. It is an interdisciplinary field that incorporates theories and methods from many fields within the broad areas of statistics, mathematics, information science, and computer science, including computer programming, artificial intelligence【人工智能】, data mining, visualization, predictive analytics, uncertainty modeling, data warehousing【数据仓库】, data compression【数据压缩】, pattern recognition【模式识别】, knowledge engineering (KE)【知识工程】, data engineering, database【数据库】, signal processing【信号处理】, probability models【概率模型】, statistical learning【统计学习】, machine learning【机器学习】, high performance computing (HPC) 【高性能计算】, and big data analysis【大数据分析】 (Figure 16.4).

Data science is the science about data especially about the massive data sets【海量数据集】 with veracity in their forms, i.e. the practice of data science is a way of extracting valuable knowledge from big data. Data science brings with its significant new research topics on data itself such as data life cycle management, data warehousing, data privacy solution, elastic data computing【弹性数据计算】, data engineering, mining big data streams, spatial-temporal nature and social aspects of big data. A practitioner of data science is called a data scientist who possesses a strong expertise in some scientific disciplines, in addition to the ability of working with various elements of computer science, mathematics, and statistics. Data science is not just restricted to big data. However, large, complex unstructured, and distributed big data sets are being collected from various sources including social media, mobile devices, wireless sensor networks, software logs, remote sensing technologies【遥感技术】, etc.【从各种不同来源收集了大量复杂的非结构化、分布式大数据集，这些数据来自于社交媒体、移动设备、无线传感网络、软件日志、遥感技术】. Data scientists employ their analytical capabilities to visualize and interpret rich data sets; collect, store, transfer, analyze, visualize, and manage large big data sets; ensure consistency of datasets; and build mathematical models of datasets【数据科学家运用他们的分析能力，从而可视化和解释丰富的数据集；收集、存储、转换、分析、可视化、管理大数据集；确保数据集的一致性；并且构建数据集的数学模型】. The efficiency of study in data science has increased due to the rapid development in big data tools related to big data storage platform and data analytic platform.

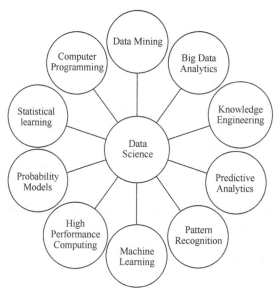

Figure 16.4 Data science

The terms "data science" and "big data" can sometimes be used interchangeably. The primary difference between the two terms is their perspective; data science begins with the data use, whereas big data begins with the data characteristics. Their formal definitions differ in more than just perspective. However, data science is intricately intertwined with big data technologies, and data-driven decision-making.

16.2 Opportunities in IoT, Cloud, and Data Science

16.2.1 Opportunities in IoT

IoT presents many opportunities and brings about new innovations. IoT will create a new

Unit 16　Internet of Things, Cloud Computing and Data Science

ecosystem by integrating the virtual world with the physical world【物联网把虚拟世界和物理世界结合起来，将创造一个新的生态系统】. Things in IoT will become context aware【上下文感知】, and the connectivity between them shall be available to all at low cost and may not be owned by private entities. IoT will become part of our everyday infrastructure【基础设施】 just like electricity, and Internet. New IoT services will be created based on real-time data. Examples of some of the IoT systems or services are: use of sensors to track RFID tags placed on products, shared video surveillance service【视频监控服务】 to monitor water levels in nearby reservoirs, RFID-tagged products and smart shelves equipped with sensors such as motion and weight detectors, implementation of Medical Body Area Network (MBAN)【医疗体域网】 to monitor heartbeat, blood pressure, temperature etc.【物联网系统或服务的实例：使用传感器来追踪放置在产品上的 RFID 标签；共享的视频监控服务来监测附近水库的水位；附着有 RFID 标签的产品和装配有传感器的智能货架，如运动和重量探测器；医疗体育网来监控心跳、血压和提问等】, using CCTV cameras and fibre optic cables【光缆】 to monitor real-time traffic conditions, and smart transportation systems.

16.2.2　Opportunities in cloud computing

Cloud computing provide opportunities for real time big data analysis because it is scalable, fault tolerant【容错】, cost effective and easy to use. Cloud Storage enables users to buy or lease storage capacity to remotely store their data and enjoy the on-demand high quality services and applications without the burden of local data storage and maintenance【云存储使用户可以购买或租用存储容量，从而远程存储他们的数据，并且不再有本地数据存储的负担和维护的需要，享受按需高质量服务和应用】. Using Cloud Computing, users can get efficient computing power. Many people say that Cloud Computing is expected to be the next generation information technology architecture. Many enterprises have their own cloud platform, such as Google, Amazon, Baidu, Alibaba. In Baidu Cloud, a user can get 2TB (terabytes) free cloud room after registering. Cloud Computing involves two major technologies: the parallel computing【并行计算】 technology, such as MapReduce and the distributed storage technology of huge amounts of data, such as GFS (Google File System)【谷歌文件系统】 and HDFS (Hadoop Distributed File System)【Hadoop 文件系统】. Apache Hadoop and Apache Spark are two popular products.

Cloud computing provides tremendous opportunities for the users. Some of them are summarized in the following subsections.

Security: Cloud provides built-in security mechanisms in individual VMs created per use. The cloud providers offer strong authentication【认证】 and verification【验证】 schemes and hire best security experts, way beyond what individual users or organizations can afford.

Energy-efficiency【能效性】: Cloud provides shared resources with efficient energy consumption mechanisms in large data centers, way beyond what manufacturers of individual systems can.

Increased storage: Cloud provides cheap storage on need basis. More storage can be provisioned in peak times and later de-provisioned when no longer required.

Cost savings: The most significant benefit to a company perhaps lies in cost savings. Whatever the type or size of a business, it aims to keep the capital and operational expenses to a minimum. With cloud, a company can have essential cost benefits by saving capital costs for server, storage, application, and infrastructure.

Reliability【可靠性】, fault-tolerance, and disaster recovery【灾难恢复】: Most cloud providers are extremely reliable and fault-tolerant in providing services than in-house IT infrastructure. They provide SLA which guarantees 99.99% availability. If a server or a network switch fails, hosted applications and services can easily be shifted to any of the available servers, so that clients are unaffected【如果一台服务器或网络交换机发生故障，托管的应用和服务可以很容易地被迁移到任何可用的服务器上，以使客户端不受影响】.

Flexibility and ease of use: Cloud is a flexible service【弹性服务】 that can be scale up, down or off based on user needs. Furthermore, it provides easy user interface【用户接口】.

Remote access【远程访问】and improved mobility: Cloud provides access to data and applications to customers no matter where they are in the world.

Automatic updates: As the business requirements changes, software needs to be updated. Cloud provides regular software updates, including security updates and hardware updates.

16.2.3　Opportunities in data science and big data

Data science is considered as the new valuable commodity for the 21st century economy. It is used to explore insights from big data and introduce new efficiencies in every industry. In order to manage big data better, new solutions and technologies are gradually proposed. NoSQL (Not only SQL) is a solution. But it can usually handle some specific area's problems. For example, MongoDB is good at data storage based documents【MongoDB 擅长基于文档的数据存储】; Redis, TTServer and HandlerSocket have very high efficiency in reading and writing key-value data【Reids,TTServer 和 HandlerSocket 在读写键值数据时很高效】; AllegroGraph can store and process images better. The impact of big data in the e-commerce is great resulting in improved customer experience, and development of better products.

The true value of data science lies in the valuable information extracted from the analysis of a group of interrelated data sets, which allow hidden principles and deep correlations to be found for trends prediction, crises management, health hazard analysis, biological and environmental research, terrorist threats detection, environmental health surveillance, search engine optimization【搜索引擎优化】, etc. However, new theories, methods, and analytic tools are still needed to help scientists and business leaders in driving valuable knowledge from the volumes of data. In the decades ahead, big data applications will play an active role in the massive global transformation towards better living standards by the integration of digital and physical infrastructure.

A good example of how big data analytics is changing the e-commerce landscape is Alibaba group, one of the world's largest e-commerce companies. AliPay【支付宝】 is Alibaba's third-party online payment platform【第三方在线支付平台】, Alibaba.com is company's English-language portal, and AliExpress【全球速卖通】, Tmall and Taobao【淘宝】 are the biggest online platforms in China. Mobile payment is also very common in daily life in China such as taking a taxi, online shopping, food delivery, and in supermarkets. The most highly using payment apps are AliPay, WeChat Payment【微信支付】, and Baidu Wallet【百度钱包】.

16.3　Challenges and Research Directions

16.3.1　Challenges and research directions in IoT

In order to realize the full potential of IoT applications, several research challenges need to be addressed. Summary of some of the research challenges are given in the following subsections.

Security issues: The key security requirements in IoT are: integrity【完整性】, privacy, authentication【认证】, access control【访问控制】, confidentiality, data consistency【数据一致性】, data usage control, trust, mobile security, secure middleware【安全中间件】, policy enforcement, and protection of data.

Trust and privacy: The devices in IoT can provide information about the user, which is counterproductive to protecting privacy. Confidence in IoT devices will depend on the protection of users' privacy.

Integration issues: The integration of IoT systems will be a challenge due to multiple platforms and devices having different operating conditions, functionalities, resolutions etc.

Scalability, data control and sharing: Handling the large amount of data produced by IoT devices is a massive challenge. Furthermore, enabling access to users' data is a barrier to large-scale adoption of IoT.

Data management: The storage and processing of the massive amount of heterogeneous data【异构数据】 produced by IoT devices is a great challenge【大量由物联网设备产生的异构数据，其存储和处理面临巨大挑战】.

Data mining: Advanced data mining techniques to mine data from IoT devices are needed.

Machine-to-Machine (M2M) communications: Novel communication protocols are needed to cope with routing【路由选择】 and congestion【拥塞】 issues.

Lack of shared infrastructure: Finding a scalable, flexible, secure and cost-efficient open source platform【开源平台】 for IoT devices is a challenge【为物联网设备找到一个可扩展的、灵活的、可靠并且成本效益高的开源平台是个挑战】.

Lack of common standards: The data produced by IoT objects, regardless of their location, must be available to users. Standardization must be done to ensure interoperability【互操作性】 of IoT devices.

Interoperability: Seamless interaction between IoT application, service provider, data provider, and data consumer is crucial to realize the true vision of IoT.

Mobility issues: Continuously providing services to mobile users is the goal of IoT.

16.3.2　Challenges and research directions in cloud computing

Cloud computing not only brings appealing benefits, but also brings new and challenging security threats. People are very afraid that storing data in the cloud will disclose their confidential information. So we hope cloud computing can offer more security measures to guarantee users' data security. In order to realize the full potential of cloud computing paradigm, several research challenges need to be addressed. These research challenges are summarized in the following subsections.

Cloud security: Security is a great concern for many users if they want to hand over important data to a third party cloud due to the shared nature of the cloud. Although cloud providers generally provide reliable and proficient security measures, clouds still need tools to detect and respond to threats and attacks.

Transition to the cloud: Moving to the cloud requires identification of the challenges and then choosing the right cloud environment.

Cloud applications: There is a variety of cloud applications ranging from simple to complex. For example, hosting simple desktop applications in a cloud; processing web data for search engines such as Google, Microsoft, and Baidu; mining unstructured data streams【非结构化数据流】 from social networking sites (SNS) 【社交网络】 such as Facebook, Twitter, Sina Weibo, Tencent Weibo, Renren, QQ, and Wechat【比如：在云上托管简单的桌面应用程序；为谷歌、微软、百度等搜索引擎处理 Web 数据；挖掘来自如 Facebook、Twitter、新浪微博、腾讯微博、人人网、QQ 以及微信等社交网络的非结构化数据流】.

QoS, service delivery, and billing: In order to get wider adoption of cloud by diverse users, clearly defined SLAs must be provided. Budgeting and assessment of the cost of on-demand cloud services is very difficult, so strong SLAs must be developed to guarantee scalability and availability.

Energy-efficiency: Energy efficiency is crucial due to limited battery of mobile devices and for reducing greenhouse gas production that results from power consumption of data centers. Furthermore, data centers that employ renewable energy sources, such as wind and solar power, and Graphics Processing Units (GPUs)【图形处理单元（器）】 must be designed.

Interoperability【互操作性】: Well-designed interoperability standards between different cloud providers are needed to provide option to users for making transition from one cloud provider to another with minimal effort and disruption of services.

16.3.3　Challenges and research directions in data science

The convergence of vast amounts of big data and massive computational power of cloud computing has given rise to increasingly widespread applications of data science. Organizations across the world have realized the growing needs of expert data scientists. However, there is confusion about what exactly is data science. It is hard to pin down a precise definition of data science because it is intricately intertwined with other aspects like mathematics, statistics, big data analytics, and data mining. However, we can say that, data science is the study of principled extraction of information and knowledge from big data.

To solve problems in data science and in order to fit the needs of the time, we must consider aspects in mathematics, statistics, and big data analytics that have particular importance to data science. Machine learning and data mining are most closely related to data science【机器学习和数据挖掘是和数据科学最紧密相关的】. Data mining is applied to perform actual extraction of knowledge from big data and machine learning is applied to discover patterns in data sets【数据挖掘用于大数据的知识提取，机器学习用于发掘数据集的模式】. Successful data scientists must be able to see all problems in the perspective of big data analytics. A data scientist must be able to systematically extract useful knowledge from data. However, it has become difficult because due to the data deluge, data is now stored in different databases, or Internet.

The following challenges need to be addressed by the researchers in order for data science to realize its potential.

Next-generation semantic data infrastructure: We need to design automated tools for searching, reusing, and integrating data.

Smart search: When multiple machines are searching online for an object or rule simultaneously, the smart search immediately updates the information. The smart search is much harder for big data using existing models and techniques. Furthermore, identifying relations in big data sets are challenging.

Algorithms: Fast algorithms are needed for high-dimensional data science processing in MapReduce.

Scalability, sparsity, and abductive modeling: The existing traditional computational tools, techniques, and architectures used in data science are largely based on conventional approaches to distributed data environments. We need to update the existing infrastructure and implement tools and techniques to accommodate the issues of scalability, sparsity【稀疏性】, and abductive modeling.

Heterogeneity, incompleteness【不完全性】, or redundancy【冗余】in data sets: The type, structure, organization, format, and granularity of data from various data sets using different analysis models is heterogeneous. Furthermore, the data sets contain high levels of duplication【重复】. Managing errors and incompleteness during data analysis and merging the results together as a single outcome is challenging.

System architecture: Building a system that processes all data science tasks is challenging. The underlying architecture must be compatible with the structure built on top of it to provide support for heterogeneous workloads.

16.3.4 Challenges and research directions in big data

Big data presents science with unprecedented challenges and problems.

Human collaboration: In spite of advancements in computational analysis, there remain some patterns where the value of human input cannot be denied. This is similar to CAPTCHAs that humans can easily detect while computer programs have a hard time finding. Crowd-sourcing【众包】is a popular method which requires support from multiple human experts. Examples are web-based encyclopedias, Wikipedia and Baidu Baike.

Privacy: Managing the privacy of big data is a great concern both from a technical and a sociological perspective, which should be addressed to realize the potential of big data. For example, most of the applications and online services require a user to share private information with the service provider, resulting in obvious privacy concerns.

Data durability【数据持久性】: Providing innovative ways to ensure long-term storage is a primary concern for big data applications. Tencent, 360 Cloud and Yun Pan (Baidu Pan), and Amazon's Glacier provide long-term cloud storage.

Data availability and storage: Big data must be stored in a way that it can be timely accessed by the users at all times. Distributed storage is still not enough for big data.

Failure tolerance and disaster recovery: One of the main issues for big data platforms is to keep operating in the event of one or more failures of components. Known ways to handle fault tolerance are redundancy【冗余】and replication【复制】. RAID (Redundant Array of Independent Disks)【磁盘阵列】is also a widely used solution. However, the scalability of these techniques is a major issue.

Transfer issues: Transfer of big data between different sites is another challenge due to speed issues of transfer of big data, traffic jam, or insecure networks.

16.4 IoT Applications

It is impossible to state all potential IoT applications. However, some of the important IoT applications are presented here (Figure 16.5).

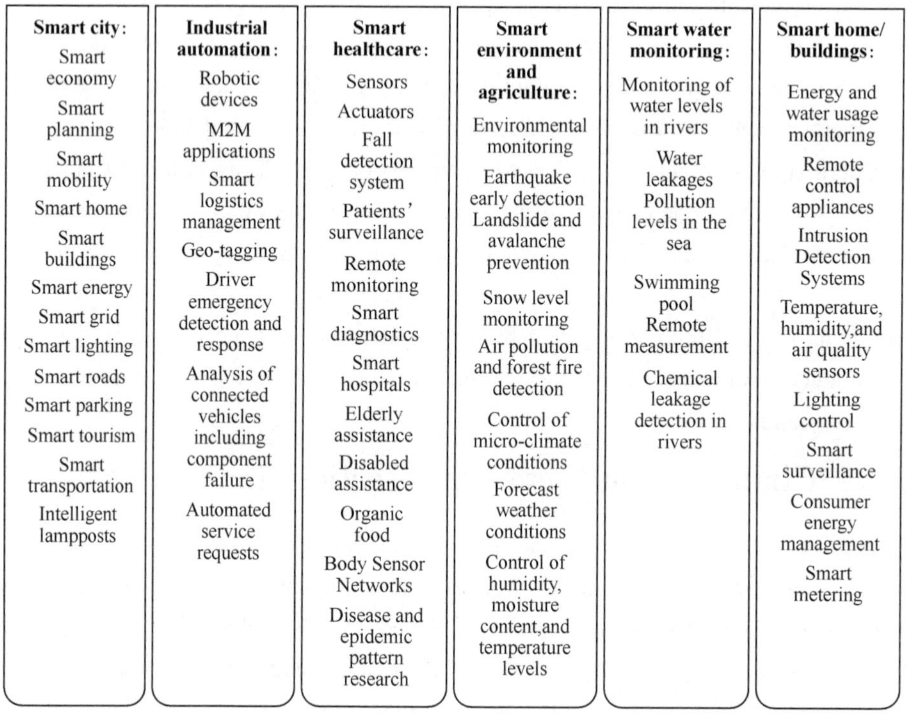

Smart city:	Industrial automation:	Smart healthcare:	Smart environment and agriculture:	Smart water monitoring:	Smart home/ buildings:
Smart economy	Robotic devices	Sensors	Environmental monitoring	Monitoring of water levels in rivers	Energy and water usage monitoring
Smart planning	M2M applications	Actuators	Earthquake early detection	Water leakages	Remote control appliances
Smart mobility	Smart logistics management	Fall detection system	Landslide and avalanche prevention	Pollution levels in the sea	Intrusion Detection Systems
Smart home	Geo-tagging	Patients' surveillance	Snow level monitoring	Swimming pool Remote measurement	Temperature, humidity, and air quality sensors
Smart buildings	Driver emergency detection and response	Remote monitoring	Air pollution and forest fire detection	Chemical leakage detection in rivers	Lighting control
Smart energy	Analysis of connected vehicles including component failure	Smart diagnostics	Control of micro-climate conditions		Smart surveillance
Smart grid	Automated service requests	Smart hospitals	Forecast weather conditions		Consumer energy management
Smart lighting		Elderly assistance	Control of humidity, moisture content, and temperature levels		Smart metering
Smart roads		Disabled assistance			
Smart parking		Organic food			
Smart tourism		Body Sensor Networks			
Smart transportation		Disease and epidemic pattern research			
Intelligent lampposts					

Figure 16.5 IoT applications

Smart cities【智慧城市】: The evolution of smart cities will have smart features, such as smart economy, smart governance, smart planning, smart transportation, smart buildings, smart citizen, smart utilities, smart mobility, smart energy, smart information communication and technology, smart parking, smart phone detection, electromagnetic field levels, smart lighting, waste management, and smart roads etc.

Industrial automation: Robotic devices are used to automatically control manufacturing, operations, production, and monitoring tasks, resulted in an obvious improvement in productivity【机器人装置用于自动化控制制造、操作、生产以及监控任务，从而大大提高生产效率】. The evolution of smart industry will have smart features, such as M2M applications, monitoring of indoor air quality, temperature, and ozone levels, etc.

Smart healthcare: Sensors and actuators in healthcare devices are used to track and monitor patient's health. The evolution of smart healthcare will have smart features, such as fall detection system for disabled or elderly people, control of medical fridges, patients' surveillance, and measurement of ultraviolet (UV) radiations, etc.

Smart environment: The evolution of smart environment will have smart features, such as earthquake early detection, landslide and avalanche prevention, snow level monitoring, air pollution, and forest fire detection etc.

Smart water: The evolution of smart water will have smart features, such as water monitoring, monitoring of water levels in rivers, water leakages, pollution levels in the sea, swimming pool remote measurement, and chemical leakage detection in rivers, etc.

Smart home and smart buildings: The life standard at home is increased by making it more convenient and easier to remotely operate and monitor home appliances, such as, air conditioner, oven, microwave, heating systems, and TV. The evolution of smart home will have smart features, such as energy and water usage monitoring, remote control appliances, and Intrusion Detection Systems (IDS)【入侵检测系统】, etc.

Smart energy and the smart grid: Energy must be produced from renewable resources such as solar energy plants. Smart grid【智慧电网】 is a transmission network which is able to efficiently route energy to the population growth with high security and quality of supply and reduced failures.

Smart transportation: Transportation will become easier and safer with Internet connections.

Organic food and water tracking: IoT can be used to secure the track of organic food and fresh water from production site to the consumer.

Smart agriculture: The evolution of smart agriculture will have smart features, control of micro-climate conditions, forecast weather conditions, and control of humidity and temperature levels, etc.

16.5 Cloud Application Service Models

Cloud providers offer services at three levels: Software as a Service (SaaS)【软件即服务】; Platform as a Service (PaaS)【平台即服务】; and Infrastructure as a Service (IaaS)【基础设施即服务】.

16.5.1 SaaS

SaaS provides access to cloud services and applications designed for customers in addition to infrastructure. SaaS applications are scalable and multi-tenant【多租户】. Some of the well-known SaaS providers are Amazon RedShift, Google Gmail, Microsoft Office 365, Salesforce, Microsoft Azure SQL, SAP Business ByDesign, and Cloud9 Analytics.

16.5.2 PaaS

PaaS provides access to platform so that users can build and manage the applications they require

【平台即服务（PaaS）提供对平台的接入，以便用户可以构建和管理他们需要的应用程序】. Some of the well-known PaaS providers are: Microsoft Azure Compute Websites, Amazon Web Services, Google App Engine, Red Hat OpenShift, Force.com, and Engine Yard.

16.5.3　IaaS

IaaS provides access to infrastructure hardware and software at an abstract or virtual level. The physical infrastructure is managed by cloud providers, while the virtual infrastructure is managed by users. Some of the well-known IaaS providers are: Microsoft Azure Compute Virtual Machine, Amazon EC2, and Google Compute Engine.

Some well-known cloud service providers in China are: Alibaba's Aliyun, Baidu, Tencent, China Mobile, China Unicom, China Telecom, and ZTE. A model of cloud platform is shown in Figure 16.6.

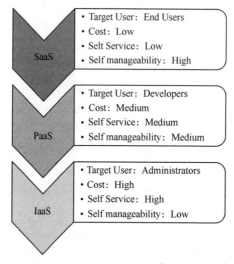

Figure 16.6　Cloud delivery types【云交付类型】

16.6　Cloud Application Deployment Models

Cloud deployment models can be classified as private clouds【私有云】, community clouds【社区云】, public clouds【公有云】, and hybrid clouds【混合云】.

16.6.1　Private cloud

Private cloud serves single organization, business or the licensee. The organization is responsible for implementation of private cloud. Private clouds are more secure but expensive.

16.6.2　Community cloud

Community cloud serves multiple organizations which have specific shared goals or requirements. A series of community members or a third-party service provider is responsible for providing the required infrastructure of community cloud. Community clouds are affordable and secure.

16.6.3 Public cloud

Public cloud serves general public over the Internet. The cloud providers (e.g. Microsoft Azure, Amazon Web Services, and Google App Engine) are responsible for providing services on pay-per-use basis. Public clouds are flexible, scalable, easy to use, and inexpensive to deploy, but the security must be well protected.

16.6.4 Hybrid cloud

Hybrid cloud is the combination of public, private, or community clouds. Hybrid clouds support elasticity, scalability, interoperability, and portability【可移植性】 on application and data【混合云支持数据和应用的弹性、可扩展性、互操作性和可移植性】.

16.7 Big Data Tools and Techniques

The ability to design, develop, and implement a big data application is directly dependent on the knowledge of the underlying architecture of the computing platform. A large number of big data tools and technologies have been developed or being developed.

Big data is often characterized by 4Vs, namely, volume【容量】, velocity【速度】, variety【多样性】, and veracity【真实性】.

Volume: It indicates a large amount of data.

Velocity: It indicates the speed of data generation, processing, transfer, and analysis.

Variety: It indicates heterogeneity in data, which can be structured, unstructured, or semi-structured【它代表数据的异构性，包括结构化、非结构化或半结构化数据】.

Veracity: It indicates level of accuracy in the data.

Apache Hadoop is an open-source fundamental framework written in Java for distributed storage and distributed processing of very big data sets on computer clusters【计算机集群】. It consists of many technologies or software packages such as MapReduce, HDFS, Apache Pig, Apache Hive, Apache HBase, Apache Mahout, Apache Avro, Apache ZooKeeper, Apache Ambari, Apache Storm and Apache Chukwa. Apache Hadoop's MapReduce and HDFS are the two primary components at the core of Hadoop and are direct implementations of Google's own MapReduce and GFS Table 16.1 shows the main function or description of Hadoop's main components.

Table 16.1 The function of Hadoop's main components

Name	Function
MapReduce	A parallel processing system of large data sets. It divides computations into two distinct steps; in the first step, the larger problem is divided into many discrete independent pieces which are fed to the map functions; this is followed by the reduce function, joining the map results back into a final product
HDFS	A distributed file system that provides high-throughput access to application
HBase	A scalable, distributed database that supports structured data storage for large tables
Hive	A data warehouse infrastructure that provides a database query interface to Apache Hadoop
Mahout	A suite of scalable machine learning and data mining library

Name	Function
Pig	A high-level data-flow language and execution framework for parallel computation
Avro	A data serialization system
ZooKeeper	A high-performance coordination service for distributed applications
Ambari	A web-based tool for provisioning, managing, and monitoring Apache Hadoop clusters
Chukwa	A data collection system for managing large distributed systems
Storm	A distributed computation framework. Data is processed in real time in Storm, while it is batched in MapReduce. Additionally, a Storm job runs indefinitely until killed, while a MapReduce job must end eventually

Spark is another framework. It is easier to use than Hadoop. And it offers more operations, such as filter, flatMap, sample, groupByKey and join, while Hadoop only has two operations: map and reduce. In order to analyze data more quickly, it supports a based-memory computing system. Except for Hadoop and Spark, many companies build their own distributed computing systems based on Hadoop, for example, Alibaba's FeiTian (like Hadoop) and Pangu (like HDFS).

16.8　Integration of IoT, Cloud Computing, and Big Data

IoT, cloud computing, and big data are conjoined (Figure 16.7). With the rapid increase in the number of IoT devices, the amount of big data produced is also increasing. Storing of such massive amounts of data is not possible with traditional storage methods on local servers. Cloud is the solution for storing, processing and analyzing big data produced by IoT devices and applications. Rather than using local storage attached to an electronic device or a computer, big data uses distributed storage technology based on cloud computing【大数据使用基于云计算的分布式存储技术，而不是使用附属电子设备或计算机上的本地存储】.

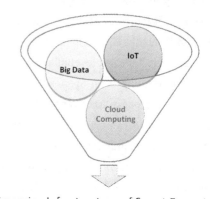

Figure 16.7　The emerging infrastructure of smarter planet【智慧地球】

Therefore, cloud is crucial as it provides services for the computation, and processing of big data produced by IoT applications and devices. For example, MapReduce is used for the processing of big data in a cloud environment, as it is designed for the processing of large amounts of datasets stored in parallel in the cluster【MapReduce 用于处理云环境中的大数据，因为它专为处理存储在并行集群中的大量数据集】. However, there is a scarcity of tools for big data processing in clouds. The actual challenges of storing and querying data cannot be solved using many well-known big data applications, such as MapReduce, Dryad, Pregel, PigLatin, MangoDB, Hbase, SimpleDB, and Cassandra. For instance, Hadoop and MapReduce have low-level data processing and management infrastructures and lack query processing strategies. Unfortunately, certain aspects of big data storing and processing in clouds are still unsolved.

The integration of IoT, cloud, and big data offers several benefits and creates opportunities for improving several interesting applications. However, at the same time, it imposes several research challenges for each application involved.

Concerns arise when some critical and sensitive IoT applications are moved to the cloud. There can be several reasons, for example, due to lack of trust in the service provider, lack of strong SLAs, and physical location of the data, being unknown to the user.

Another big challenge in the integration of IoT, cloud, and big data is the heterogeneous nature of big data coming from different devices having varied platforms, operating systems, architectures, and standards.

Therefore, integrating IoT, cloud and big data has advantages, as well as some concerns. Although, cloud has overcome some of these challenges, such as the reliability of IoT devices is increased in cloud by allowing offloading of heavy tasks resulting in increased devices' battery duration. But still, there are some risks in moving to a cloud.

It is estimated that, in the next few years, there will be a massive increase in the number of connected IoT devices. IoT will be one of the main sources of big data, and cloud will enable to store it for long time and to perform complex analyses on it. The ubiquity of mobile devices and sensor pervasiveness, certainly call for scalable computing platforms (every day 2.5 quintillion bytes of data are produced). The handling of this large amount of data is a critical challenge, since already existing data management services are not enough. Development of some cloud based big data techniques are still under investigation. For instance, big data summarization based on semantic extraction【语义提取】 is still a challenging task for existing cloud applications. Therefore, after NoSQL, some alternative database technologies for big data are offered, such as timeseries【时间序列】, key-value, document store, wide column stores, and graph databases【因此，继 NoSQL 之后，出现了一些针对大数据可供选择的数据库技术，比如时序数据库、键值数据库、面向文档的数据库、列存储数据库、以及图数据库】. However, there is no single perfect solution to manage the big data on clouds produced by all the IoT devices. Moreover, security, privacy and data integrity are very crucial factors for IoT data on clouds.

16.9 References and Recommended Readings

For more details about the subjects discussed in this chapter, the following books are recommended:
- Acharjya D P, Dehuri S , Sanyal S. Computational Intelligence for Big Data Analysis. Frontier Advances and Applications, Adaptation, Learning, and Optimization, Volume 19, Springer,

2015.
- Bessis N , Dobre C. Big Data and Internet of Things. A Roadmap for Smart Environments, Studies in Computational Intelligence, Volume 546, Springer, 2014.
- Borgia E. The Internet of Things vision: Key features, applications and open issues. Computer Communications, Volume 54, 1–31, 2014.
- Chen M, Mao S, Zhang Y , Leung VCM. Big Data. Related Technologies. Challenges and Future Prospects, SpringerBriefs in Computer Science, Springer, 2014.
- Hassanien A E, Azar A T, Snasel V, etc. Big Data in Complex Systems: Challenges and Opportunities. Studies in Big Data, Volume 9, Springer, 2015.
- Hurwitz J, Nugent A, Halper F etc. Big Data For Dummies. Hoboken, NJ: John Wiley & Sons, 2013.
- Jagadish H V. Big Data and Science: Myths and Reality. Big Data Research, 2, 49-52,2014.
- Li K C, Jiang H, Yang L T etc. Big Data: Algorithms, Analytics, and Applications. Big Data Series, Chapman & Hall/CRC, 2015.
- Loshin D. Big Data Analytics: From Strategic Planning to Enterprise Integration with Tools, Techniques, NoSQL, and Graph. Waltham, MA: Elsevier, 2013.
- Machiraju S , Gaurav S. Hardening Azure Applications. Suren Machiraju and Suraj Gaurav, 2015.
- Marinescu DC. Cloud Computing: Theory and Practice. Waltham, USA: Elsevier, 2013.
- Saidi A A, Fleischer R, Maamar Z etc. Intelligent Cloud Computing. First International Conference, 2014.
- Smith I G. The Internet of Things. 2012 New Horizons, Halifax, USA: IERC, 2012.
- Underdahl B. The Internet of Things For Dummies. Hoboken, NJ: John Wiley & Sons, 2013.
- Vermesan O , Friess P. Internet of Things. Converging Technologies for Smart Environments and Integrated Ecosystems, River Publishers, 2013.

16.10 Summary

- The transformation from digital to real-time intelligence has made IoT, cloud, and data science indispensable to the very nature of work. These technologies have revolutionized our globe and are considered as complementary as they have enhanced each other capacities and capabilities.
- IoT, sometimes called Internet of Everything (IoE), represents a paradigm in which any physical thing can become a computer that is connected to a dynamic and self-configuring global network.
- Things included in IoT are much more than everyday objects, vehicles, electronic equipment, smart phones, mobile devices, utility meters, cameras, Bluetooth, Wi-Fi, Satellite, Ethernet, global positioning systems (GPS), sensors, RFID, wireless sensor networks (WSNs), Machine-to-Machine (M2M) communication, 2G/3G/4G/5G, IP, industrial systems, agricultural systems etc.

- The technologies that make intelligent IoT applications possible are considered as the enabling technologies of IoT.
- Cloud computing is a model in which Information Technology (IT) resources and services are abstracted from the underlying infrastructure and accessed on demand by the customers based on service-level agreements (SLAs) established through negotiations between the cloud service providers (CSPs) and consumers. The cloud service can be hosted on-site or off-site such as Tencent cloud, Alibaba's Aliyun, Xiaomi's Mi Cloud, Apple's iCloud, Microsoft's SkyDrive and Samsung's S-Cloud.
- Some major cloud providers are Amazon Web Services (AWS), Microsoft Windows Azure, Rackspace, and Baidu Cloud.
- Data science is the study and practice of extracting knowledge from data and finding valuable insights in data to help make valuable decisions. It is an interdisciplinary field that incorporates theories and methods from many fields within the broad areas of statistics, mathematics, computer science, and information science.
- IoT presents many opportunities and brings about new innovations. Examples of some of the IoT systems or services are: use of sensors to track RFID tags placed on products, shared video surveillance service to monitor water levels in nearby reservoirs, RFID-tagged products and smart shelves equipped with sensors such as motion and weight detectors, implementation of Medical Body Area Network (MBAN) to monitor heartbeat, blood pressure, temperature etc., using CCTV cameras and fibre optic cables to monitor real-time traffic conditions, and smart transportation systems.
- Cloud Computing involves two major technologies: the parallel computing technology, such as MapReduce and the distributed storage technology of huge amounts of data, such as GFS (Google File System) and HDFS (Hadoop Distributed File System). Apache Hadoop and Apache Spark are two popular products.
- With the rapid increase in the number of IoT devices, the amount of big data produced is also increasing. Cloud is crucial as it provides services for the computation, and processing of big data produced by IoT applications and devices in IoT. However, unfortunately, there is a scarcity of tools for big data processing in clouds. Development of some cloud based big data techniques are still under investigation.

16.11   Practice Set

1. What is data science?

2. According to some people there is a fundamental difference between Computer Science (CS) and Information Technology (IT). Might there be a fundamental difference between data science and big data? If yes, then would that difference help/harm the understanding of "scientific" challenges and principles of data science?

3. What are the 4 Vs of big data?

4. What is meant by the term "data deluge" or "petabyte age"?

5. Compare and contrast the deployment models of cloud.

6. Mobile devices could benefit from cloud computing; if you agree with this statement, explain with reasons, and otherwise provide arguments supporting the contrary.

7. Compare and contrast the three cloud computing delivery models, SaaS, PaaS, and IaaS, from the point of view of end-user. Discuss the merits and demerits of each model.

8. An IT company decides to offer free access to a public cloud dedicated to higher education. Which one of the three cloud computing delivery models, SaaS, PaaS, or IaaS, should it adopt, and why?

9. What is the relation between IoT, cloud computing, and big data?

10. What are the potential applications of the Internet of Things?

11. What is the vision of the IoT?

12. What are the most challenging technical issues in IoT, cloud, and big data that must be resolved?

13. What is a smart ecosystem?

14. Identify present and future enabling technologies that make the IoT concept feasible.

15. Define application scenarios for IoT technology and identify industry needs for IoT applications.

16. Identify challenges for the IoT.

17. What are the challenges in integrating big data, cloud, and IoT?

18. Write briefly about some of the main components of Hadoop.

19. HDFS is a _____ capability of big data.
    a. storage and management   b. processing
    c. database   d. programming

20. _____ refers to the different type of data coming from different sources.
    a. Volume   b. Variety   c. Veracity   d. Velocity